Basic Principles of Polymer Chemistry

Archana Garg
M.Sc., Ph.D.

Professor of Chemistry
M.D. University
Rohtak, Haryana

ATLANTIC
PUBLISHERS & DISTRIBUTORS (P) LTD

Published by

ATLANTIC

PUBLISHERS & DISTRIBUTORS (P) LTD

7/22, Ansari Road, Darya Ganj, New Delhi-110002
Phones : +91-11-40775252, 40775214, 23273880, 23275880
Fax : +91-11-23285873
Web : www.atlanticbooks.com
E-mail : orders@atlanticbooks.com

Branch Office: Chennai
Phones : +91-44-48531784, 28291383
E-mail : chennai@atlanticbooks.com

Printed & bound in India by Atlantic Print Services

Dedicated to

My father
Late Prof. Dr. A.C. Mittal
who encouraged me to take up Chemistry,
and to my mother
Shrimati Krishna Mittal
who has always patted my back for
every little achievement

"An ocean of electric Energy
Formlessly formed its strange wave-particles
Constructing by their dance this solid scheme,
Its mightiness in the atom shut to rest...."

—*Savitri* (Sri Aurobindo)

Preface

Polymers are high molecular weight compounds formed by a combination of low molecular weight compounds known as monomers, more precisely known as polymerization. They are described considering several factors like by the extent of their degree of polymerization, molar mass distribution in a polymer sample, tacticity, extent of cross linking and other properties. To a layman, polymers are known by several common names, like plastics, resins etc.

This book is an effort to present a clear understanding of the basics of the subject of polymers for the students of undergraduate and postgraduate levels. The book is divided into four parts; these four parts are further divided into eight chapters. The first part introduces the genesis of polymers. The second part has four chapters on polymerization. It deals with the synthesis of chain polymers, step polymers and copolymers and kinetics of all these syntheses. The third part of the book is the characterization of polymers, which is one of the most important aspects of polymers. This part has a chapter in which some of the important methods for to determine molecular weight is given. In the fourth part there are two chapters; the structural aspects of polymers viz. the glass transition temperature is discussed in seventh chapter and the eighth chapter deals with the thermodynamic and other aspects of the polymer solution.

No single book could be a comprehensive one to serve all the needs of a student and cover all the topics of this field. But nevertheless, I hope this book will serve to develop a clear understanding of the basic aspects of polymers for students.

I would like to thank Maharshi Dayanand University for granting me the time and facilities to accomplish the project of writing this book. A special thanks to Dr. V K Sharma (HOD Chemistry, MDU) for being an encouraging shoulder for the project. Many a thanks to my dear colleagues Dr. Sapna Garg and Dr. Rajni Arora for innumerable things for which there isn't enough space to enlist. My gratitude to my husband, Shri Vinod K Garg, who continuously stood by my side, taking up the responsibilities at home, while, my academic work and health issues kept me occupied.

Many people have contributed directly or indirectly to this book, which I might not even realize but salute them with a note of thanks as they stood by me humbly without expectations. One person whom I can't thank enough is Dr. Pratyoosh Shukla who has helped, encouraged and contributed to this project in infinite ways and will always have my regard and gratitude. I shall remain ever grateful to Shri K Srinivasulu, my guide and mentor. I thankfully appreciate my son, Pranjal, who looked into the technical and lexical aspect, and the semantics of this book, pulled out a lot of his precious time during his doctoral thesis and helped me push the project to its completion. A thankful note is also to the Atlantic Publishers & Distributors (P) Ltd., who, promptly worked on my book draft and its publication.

24.garg@gmail.com **Archana Garg**

Contents

Part – A
Introduction

1 The Genesis of Polymers

1.1 Introduction and Basic Concepts

Polymers stand out among other compounds because of their high molecular weight. They are widespread and have varied applications in today's society. Polymers such as proteins, nucleic acids, polysaccharides, blood, cells, etc. are the basis of living organisms. Following are some of the applications of polymers in our day-to-day life, e.g. they are used in construction materials like concrete, insulation, and wooden and composite beams; in packaging of food, breakable items and others; for home furnishings carpets, pillows, curtains, coatings, baskets and buckets, water pipes, window glass, chair, bed, kitchenware; for manufacturing transport vehicles like aircraft, automobiles, ships, and trucks; medical devices like artificial organs, ligaments, pacemakers, contact lenses, bone cement, variety of glasses for eyes, etc.; in information technology devices and materials such as telephones, televisions, computers, compact discs (CDs), newspapers, optical fibres, and cell phones, etc.; for objects of sports and entertainment like various toys, plastic models, skipping ropes and so on; for making various types of cloth like textured cloth and fire-proof cloth. Similarly, in many other walks of life, we encounter continuously the 'polymer-footprint'.

The size of the polymers makes them very important. Many polymers are made from low cost and abundantly available materials into a variety of products. The polymerized products are sometimes very expensive compared to the raw material used in their synthesis. Synthesized polymers often have desirable physical and chemical properties. There are a large number of polymers which have high strength, high viscosity, elasticity, glass transition temperature and several other properties. It is because of such unique properties that polymers are so useful for mankind. Some polymers are stronger on a weight basis than steel and do not degrade quickly and are also resistant to rust and corrosion. Most of the research on polymers were carried out during the First World War and has continued since then. Today, polymers occupy the most important place in sustaining and carrying out life on the earth. No other material can match polymers for its incredible variety and diversity of the applications, performance, chemical and physical properties which are found in polymers. Polymers have completely changed the present-day human civilization and culture.

Today we have realized that ordinary covalent bonds are the intra-molecular forces that keep the polymer molecules intact. In addition, the usual types of inter-molecular forces like hydrogen bonds, dipole-dipole interactions, and London forces, etc. hold these molecules together in the bulk state. The most remarkable thing about these molecules is their size and indeed, that feature is noteworthy.

The discovery and development of polymer science is credited to the outcome of the dedicated work of H. Staudinger, G. Natta, K. Ziegler, P.J. Flory, B. Merrifield, Pierre-Gilles de Gennes, H. Shirakawa, A.G. MacDiarmid, H.F. Mark, S.R. Palit, W.H. Carothers and the efforts of many more. Their endeavour has brought to us the deep knowledge of polymers and their properties and useful applications in numerous fields. Polymer research is a field that has a maximum number of Nobel prizes to its credit. H. Staudinger in 1953, G. Natta and K. Ziegler shared in 1963, P.J. Flory in 1974, B. Merrifield in 1984, A.G. MacDiarmid, H. Shirakawa and A.J. Heeger shared in 2000, J.B. Fenn and K. Tanaka together and K.T. Wuthrich shared Nobel prizes in 2002 for their work on polymers.

Polymers are macro-molecules built up by linking together of a large number of much smaller molecules known as monomers. There may be a hundred, a thousand, tens of thousands, or more monomer molecules linked together in a polymer molecule. Unlike the molecular weight of small molecules like water and benzene, known as low molecular weight materials polymers have a high molecular weight which varies from 10,000 g/mol to 10^7 g/mol.

The molecules whose molecular weight lies between these two are often referred to as oligomers. However, there is no hard and fast rule for this.

In order to have an idea of the length of a polymer molecule, we convert the size of a polyethylene molecule which roughly contains 50,000 methylene groups, to a common man's conceivable scale, i.e. if the length of one ethylene molecule is considered to be 1 cm, the polyethylene chain would be about 250 metres long. This makes L/D ratio (length/diameter) in polymer molecules very high as compared to simple molecules. This high L/D ratio makes polymers a different class and provides them different physical and chemical properties. It is due to a long chain that polymers have inter-molecular and intra-molecular entanglement which is affected by the prevailing thermo-chemical conditions. The properties like high melting point, high softening temperatures, and high melt viscosity also arise due to high length and entanglement of the chains in addition to other factors.

In India, polymers have been used as medicine and food since immemorial times. There are several medicines used, are natural polymers, e.g. *Gaund Katira*. Its cooling effect on the body is well known to the traditional folk. The fried *Gond* is eaten in winters for keeping the body warm, which is a natural polymer. Its aqueous solution is used as an adhesive too.

Berzelius was the first to introduce the term 'Polymer' in 1827. But it was H. Staudinger, a German scientist, who in 1920, for the first time used the term Polymer for molecules with high molecular weight (at least a few thousand). Another synonymous term 'Macro-molecules' appeared for polymers, but only in the context of polymers of biological origin. Staudinger also gave long chain structural formulas for natural rubber, cellulosic such as cellulose nitrate, and polystyrene. This established the basis of macro-molecular or chain molecular concept for polymers. A long sequence of simple chemical units interlinked with covalent bonds was recognized as the basic structural feature for all polymer molecules.

Carothers did his pioneer work in the area of polymer synthesis by polycondensation and synthesized the first linear macromolecules, viz. polyesters and polyamides. The unusual properties of polymers and hence their varied applications motivated many researchers towards synthesizing of new polymers. And within thirty years a stage was reached when the synthesis of polymers of required property became conceivable. This was done by physical and chemical modifications in functional groups, compounding and reinforcing polymers, making polymer blends, polymer alloys, sometimes the reaction mixture was irradiated and sometimes thermal treatments and mechanical treatments. By then, a stage is reached that the tailor-made polymers could be synthesized. A new branch named "Molecular Engineering" was developed which had its starting point in the realization of the long-chain molecular systems.

1.2 What are Polymers

For those who are new to the field of polymers, the question is "what is the difference between plastic and a polymer and are they the two names for one and the same thing?" The answer would be that plastic is a type of a polymeric material, whereas polymer is a type of molecule, which is extremely large in size. The word plastic comes from *plastikosa*, a Greek word, meaning malleable. The word polymer also comes from Greek, where 'poly' means 'many' and 'mer' means 'of many parts'.

1.3 How are Polymers Made

Most polymers have an organic origin and are formed of hydrocarbon molecules. However, inorganic polymers also exist but they are lesser in number. Polymers are made by the polymerization of monomers. All the substances cannot be polymerized for it is necessary for the monomer to have some functional characteristics in it.

Repeat unit in a polymer chain ("unit cell") is called a mer. Small molecules from which polymer can be synthesized is a monomer. A single mer is commonly called a monomer. When all mers in the reaction mixture are the same, the resulting big molecule is called a homopolymer. When more than one type of mers is present, the molecule is called a copolymer. A mer unit that has two active bonds to connect with other mers is called bifunctional. They result in a log-chain polymer with or without branching. A mer unit that has three active bonds to connect with other mers is called trifunctional. They form three-dimensional molecular network structures. It needs to be emphasized again here that all substances couldn't be polymerized. If the chemical substance is bifunctional (which could be because of the presence of either an unsaturated bond or two reactive functional groups) then it can act as a monomer and will result in a polymer.

$$\underset{\text{Ethylene}}{H_2-C=C-H_2} \rightarrow \underset{\text{Polyethylene}}{(CH_2-CH_2)_n}$$

$$\underset{\text{Ethylene glycol}}{-HO-CH_2-CH_2-OH} \rightarrow \underset{\text{Polyethylene glyol}}{HO\,(CH_2-CH_2-O)H}$$

It has been my experience that the need of bi-functionality of a monomer is a must for the polymer to synthesize, which is usually not well understood by the students. So, let us understand this first by taking the following example.

Let's take two compounds, acetic acid and ethyl alcohol (each possessing one functional group, viz. —COOH and —OH). Their reaction will complete in one step resulting in the formation of ethyl acetate.

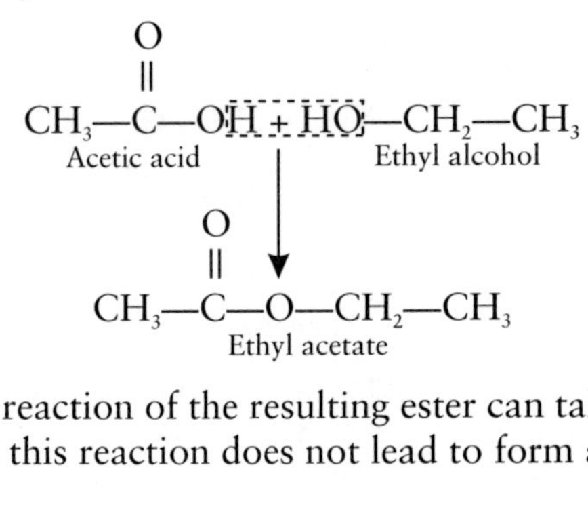

No further reaction of the resulting ester can take place and we can see that this reaction does not lead to form a polymer.

Now we carry out another reaction between acetic acid and ethylene glycol. This reaction takes place in the following two steps.

1st step

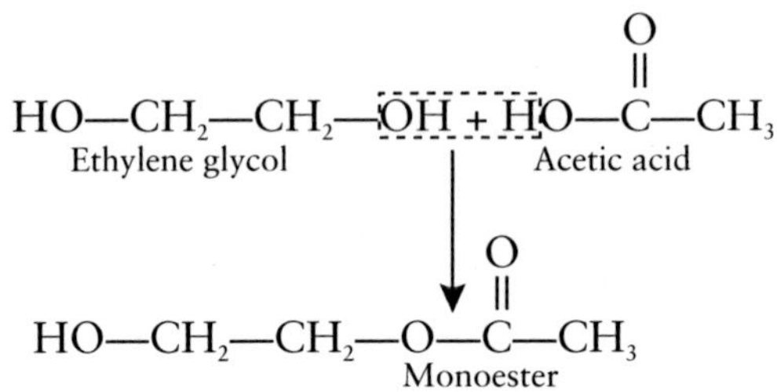

2nd step

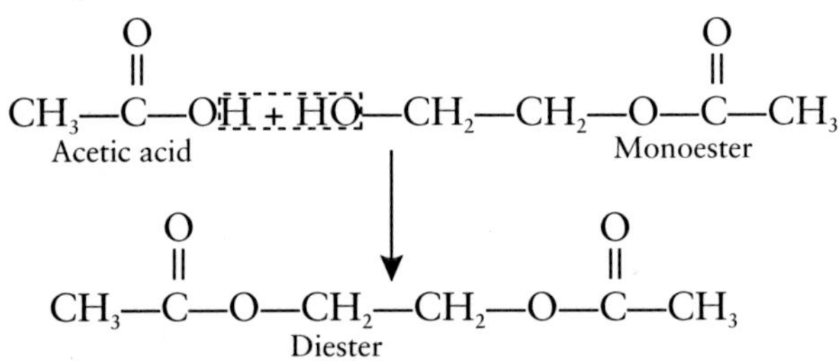

Here, at this stage, the reaction stops, as there is no functional end group left to react to form a polymer.

In the third case, we consider a reaction between ethylene glycol (a diol) and malonic acid (a dibasic acid).

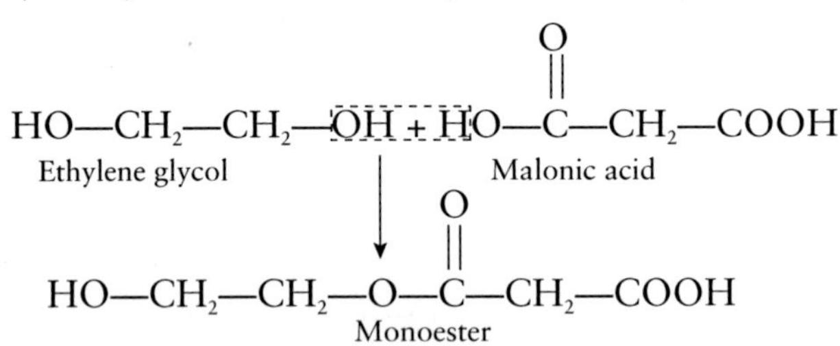

is a monoester with active ends. One end has —OH group and another end has —COOH group, which are active. This monoester is able to react with another molecule of malonic acid and ethylene glycol.

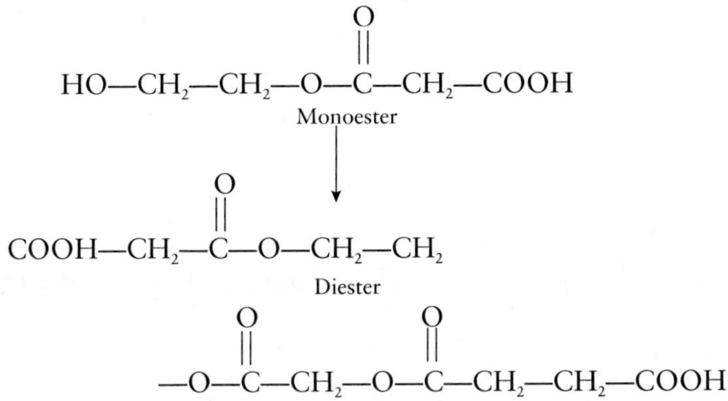

This process keeps on repeating to form long chain polyester. The length of the chain will depend on the availability of the monomers for the reaction in addition to other factors. Thus, the reaction between two bi-functional molecules leads to a linear chain polymer. Further, if we add a monomer with a tri-functional group, it may form either a branched polymeric chain or a three-dimensional network structure.

1.4 Classification of Polymers

There are various ways to classify polymers based on various considerations like natural or synthetic, type of polymer, type of chain, etc.

Polymers are generally classified on the basis of:

I. Source: Natural, semi-synthetic and synthetic polymer.

II. Physical and chemical structures: Linear, branched and cross-linked polymer, crystalline and amorphous polymer (on the basis of arrangement of chains), homochain and heterochain polymer, Isotactic, syndiotactic and atactic polymer, Organic, elemento-organic and inorganic polymer, monodispersed and polydispersed, Homopolymers and copolymers.

III. Preparation method: Addition and condensation polymer.

IV. Physical properties (molecular forces): Thermoplasts and thermosets, fibers, plastics and elastomers, charged and uncharged polymers.

V. Applications: Adhesives, coatings, films and fibres.

Following are the classifications:

1. Natural and synthetic polymer
2. Linear, branched and cross-linked polymer
3. Crystalline and amorphous polymer (on the basis of arrangement of chains)
4. Homochain and heterochain polymer
5. Isotactic, syndiotactic and atactic polymer
6. Organic, elemento organic and inorganic polymer
7. Monodispersed and polydispersed
8. Homopolymers and copolymers
9. Addition and condensation polymer
10. Thermoplasts and thermosets
11. Fibres, plastics and elastomers
12. Charged and uncharged polymer
13. Adhesive, coatings and films, fibres.

1. *Natural, Semi-synthetic and Synthetic Polymer*

All the polymers obtained from natural resources like cotton, wool, silk, asbestos, proteins, cellulose, resins, rubber, nucleic acids (RNA, DNA), etc. are termed as a natural polymer. Sometimes they are also termed as Biopolymers or Biological Macromolecules. All man-made polymers are synthetic polymers, e.g. polyethylene, Nylone 66, polyamides, silicone resins, etc. Semi-synthetic polymers are those which are obtained by subjecting natural polymers to chemical processes so as to rearrange the structure of existing molecules, especially large molecules containing carbon, e.g. formation of cellulose acetate (rayon) and cellulose nitrate from naturally occurring cellulose, vulcanized rubber.

2. *Linear, Branched and Cross-linked Polymer*

If a polymer has a long chain then it is known as a linear polymer. The polymers resulted by branched monomer are termed as branched polymers. And the polymers resulted from polyfunctional monomer have network structure and are termed as cross-linked polymers.

3. *Crystalline and Amorphous Polymer*

This classification depends on the morphological behaviour of polymer whether their structure is crystalline or amorphous. Following are the factors that cause the polymer to become crystalline:

- Cooling rate
- Chain complexity and regularity
- Side group size
- Tacticity
- Cross-linking
- Branching

However, in any of the polymers, 100% crytallinity can never be obtained. Crytallinity affects several properties of the polymer like hardness, elasticity, etc. The solid states of the crystalline polymers have a long-range order of molecular chains while the amorphous polymers have a short-range order of repeating units. In crystalline polymers, the chains are aligned, while in an amorphous polymer the chains lie random. The melting of both types of polymers increases disorder. However, there are some polymers which have alternate ordered and disordered regions. They are known as liquid crystalline polymers. The molecules with an ordered structure are known as crystalline polymer, while the molecules with disordered structure are known as amorphous polymer. Solids liquid crystalline polymers melt down to form fluids in which much of the molecular order is retained within a certain range of temperature.

4. *Homochain and Heterochain Polymer*

If the main chain of a polymer is made up of linkages between the atoms of the same element, e.g. carbon in an organic polymer, silicon in an inorganic polymer then they are termed as homochain polymer. But, if the main chain consists of linkages between different atoms, e.g. carbon and oxygen, carbon, nitrogen and oxygen, etc. then they are termed as heterochain polymer.

5. Isotactic, Syndiotactic and Atactic Polymer

Polymers can have stereoisomerism where the atoms are linked together in the same order, but can have different spatial arrangements. This can be of three types.

(i) Isotactic polymers are those in which all side groups R are on the same side of the chain of tertiary C-atoms.

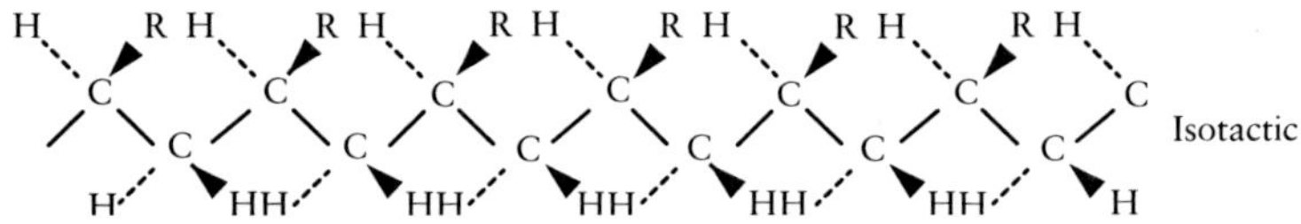

Fig 1.1 (a)

(ii) Syndiotactic polymers are those of which the chains consist of regular series of monomer units. And side groups R are present on every tertiary C-atom of the chain. These side groups R possess opposite steric configuration.

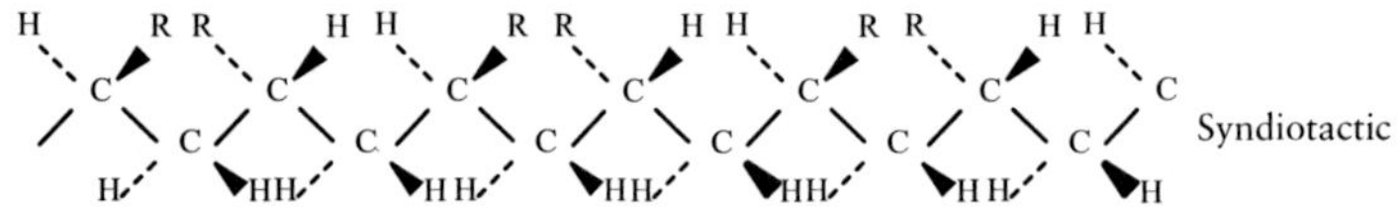

Fig 1.1 (b)

(iii) Atactic polymer lacks any regularity of side group R in the chain in the distribution of steric configuration of monomer units.

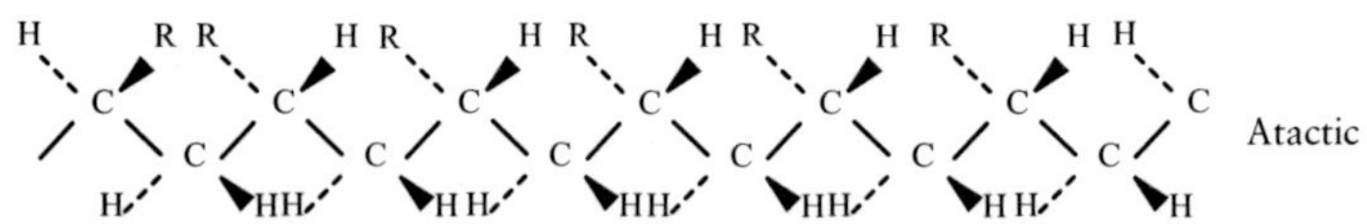

Fig 1.1 (c)

6. *Organic, Element Organic and Inorganic Polymer*

Organic polymers have carbon backbone. They may also have oxygen, nitrogen, sulfur and halogen, etc. in the side chain in addition to hydrogen.

Element-organic is macromolecules in which the backbone has carbon and any of the following elements—beryllium, boron, silicon, phosphorous, aluminium, titanium, iron, cobalt, copper, germanium, arsenic, tin, tellurium, mercury, thallium and lead, etc. These elements could be introduced in the backbone because of the similarity in their physical and/or chemical properties. Inorganic linear polymers contain Si, N, P, B with oxygen. Inorganic polymers with even three or more inorganic elements are quite frequent. Some of the examples are polysilanes, polysiloxanes, polyphosphazenes, polyboron oxides, etc.

7. *Monodispersed and Polydispersed Polymers*

During the synthesis of a polymer, the process of the polymer chain growth is governed by random events. With the result varied polymer chain lengths are obtained. Hence, in contrast to low molecular weight compounds, polymers cannot be characterized by single molecular weight rather statistical average molecular weight is taken.

When all the molecules of a polymer are of the same size, it is called monodispersed. In this case, the number average molecular weight of the polymer is equal to its weight average molecular weight, i.e. the $\bar{M}_w / \bar{M}_n$ ratio is equal to 1.

If there is a greater distribution of molecular sizes so that polymer chains of different molecular weights are obtained, the polymer is called polydispersed. In this case, there will be a greater disparity between the two averages and polymeric disparity given by $\bar{M}_w / \bar{M}_n$ will is greater than 1.

Generally in synthetic polymers and many naturally occurring polymers are polydispersed.

8. *Homopolymers and Copolymers*

The polymers made up of only one type of monomers are known as homopolymers. While the polymers made from more than one type of monomers are known as copolymers. The main characteristic feature of copolymers is that they have characteristics of both the type of monomers. Hence, instead of searching for new polymer, one can choose various ratios of both types of monomers to gain a required property of a polymer.

9. *Addition and Condensation Polymer*

The polymers synthesized by addition polymerization are addition polymer and the polymers synthesized by condensation polymerization are termed as condensation polymers. During condensation polymerization, there may or may not liberate a small molecule while in addition polymerization no such liberation occurs at all.

10. *Thermoplasts and Thermosets*

One of the oldest ways of the classification of a polymer is based on their response to heat is the oldest way. However, this is the most useful classification for engineering purposes. The linear polymers that take new shapes by repeated application of heat and pressure are thermoplasts. They are composed of linear monomers or monomers having moderate branching. They are soluble in solvents but do not have appreciable thermal resistance properties. Vinyl, cellulose derivatives, polyethylene and polypropylene fall into this category. While cross-linked polymer flows or melt only once and are known as thermosets. These are the polymers, which decompose when heated and hence cannot be reshaped. Such polymers have a complex 3-D network (cross-linked or branched). They are generally insoluble in solvents and have good heat resistance property. Thermosetting polymers include phenol-formaldehyde, urea-aldehyde, silicones and allyls.

11. *Fibres, Plastics and Elastomers*

This classification is based on the alignment of molecules, which in turn depends on the molecular structure. This depends on how the molecules are hooked together.

Fibres: The molecules with regular structure are aligned very closely using secondary intermolecular bonding forces.

Because of this close alignment, a fibrous structure is obtained. On this basis, fibres can be defined as linear polymers with high symmetry and strong intermolecular forces. Polar groups are responsible for strong intermolecular forces. Fibres have high modulous, high tensile strength and moderate extensibility. They are crystalline in nature, but can form thread-like structures. Examples of fibres are polyamides, polyesters and polyurethanes, protein derivatives, etc.

Elastomers: They are amorphous liquids and have glass transition temperature below room temperature. They have an irregular structure with weak intermolecular forces of attraction. The elastomer chains are very flexible. In the absence of applied (tensile) stress, the elastomer molecules become coiled and extensible. Examples of elastomers are butadiene, butadiene co-polymers and their derivatives, polycondensation products, silicones and thiokols.

Plastics: Plastics are glassy polymers, and from the structural point of view can be placed between fibres and elastomers. The term 'Plastic' should not be confused with the common word 'plastic' that is applied to any material that changes shape permanently when pushed beyond its tensile strength.

12. *Charged and Uncharged Polymer*

The polymer containing free functional groups are known as charged polymer, e.g. polyacrylic acid (anionic polymer), polyethylimine (cationic polymer). The charged polymers which are soluble in water are called polyelectrolytes while water insoluble charged polymers are called ionomers.

13. *Adhesive, Coatings and Films and Fibres*

Adhesives, coatings and films and fibres fall under the category of application polymers.

Adhesives are resinous and water resistant polymers, e.g. cellulose adhesives, vinyls and natural products such as starch, dextrins, proteins and natural rubber, etc.

Coatings and films are polymers that are used for protective and decorative purposes, e.g. films of polyethylene and cellulosic types.

Fibres are formed by extrusion or spinning of linear monomer molecules into thin sections of diameter in the range of 10-50 microns. Fibres have excellent tensile strength, creep and resilience. Cotton, wool, etc. are natural fibres.

1.5 Nomenclature of Polymer Chains

Up till now we have a large number of known polymers (up till 2014 Chemical Abstracts have more than 60,000 polymers) that have a variety of structures. The nomenclatures given for polymers are based either on the structure of the polymer or the source of the polymer (the monomers used in its synthesis). However, sometimes trade names and sometimes a semi-systematic set of trivial names are used. It is very common to use trade names for commercial synthetic polymers. Biopolymers are often named according to their function, preparation, or characterization. In order to avoid any ambiguity, International Union of Pure and Applied Chemistry (IUPAC) and the American Chemical Society (ACS) in the 1970s have jointly recommended a system of nomenclature based on the structure of the monomer or repeat unit in the polymer.

According to the IUPAC nomenclature, if a single monomer is used for the synthesis of the polymer, e.g. simple vinyl polymers, it is named by attaching the prefix 'poly' to the monomer name, e.g. polystyrene, polyethylene, and polypropylene. But if the monomer name consists of more than one word or is preceded by a letter or number, the monomer is enclosed in parentheses preceded by the prefix 'poly'. For example, the polymer obtained from the polymerization of 4-chlorostyrene is poly(4-chlorostyrene) and that from vinyl acetate is poly(vinyl acetate).

Tacticity may be noted by prefixing the letter 'i' for (isotactic) and 's' for (syndiotactic) is placed before 'poly' as in i-polystyrene.

Geometric and structural isomers may be indicated by using the appropriate prefixes, cis or trans and 1,2- or 1,4-, before 'poly', as in trans-1,4-poly(1,3-butadiene).

The rules of IUPAC nomenclature for condensation polymers, which are a product of two different monomers, are somewhat more complicated than for vinyl polymers. They are usually named according to the functional group of the repeat unit. For example, the most important commercial nylon, commonly called nylon-6,6 (66 or 6/6), is called poly(hexamethyleneadipamide). Similarly, the aliphatic nylon obtained by the polyamidation of hexamethylenediamine and sebacic acid is named poly(hexamethylenesebacamide).

But in some cases, "common" names are used exclusively in place of chemically correct IUPAC nomenclature. For example, the IUPAC nomenclature for condensation polymer, a thermoplastic made from phosgene and bisphenol-A is 2,2-bis(4-hydroxyphenyl)propane. But its common name bisphenol-A polycarbonate is much in use. The prefix bisphenol-A is placed before polycarbonate to distinguish it from other polycarbonates that are synthesized from other bisphenol monomers like tetramethylbisphenol-A.

For many years, the International Union of Pure and Applied Chemistry (IUPAC) and the American Chemical Society (ACS) have developed a detailed, structure-based nomenclature for polymers. In addition to this, the American Society for Testing and Materials (ASTM) developed an industrial standard for Testing and Materials (ASTM D-4000) for specifying specific commercial grades of reinforced and non-reinforced plastics.

The copolymers are named by IUPAC, by incorporating an italicized connective term between the names of monomers contained within parentheses or between two or more polymer names. Six such copolymer classes have been categorized as follows:

1. Unspecified (-co-added) Poly[styrene-*co*-(methyl methacrylate)]
2. Statistical (-stat-added) Poly(styrene-*stat*-butadiene)
3. Random (-ran-added) Poly[ethylene-*ran*-(vinyl acetate)]
4. Alternating (-alt-added) Poly[styrene-*alt*-(maleic anhydride)]
5. Block (-block-added) Poly[styrene-*block*-polybutadiene]
6. Graft (-graft-added) Poly[butadiene-*graft*-polystyrene]

Note: A statistical polymer has a sequential distribution of the monomeric units and obeys the statistical laws. In the case of a random copolymer, the probability of finding a given monomeric unit at any site in the chain is independent of the neighbouring units in that position.

1.6 Isomerism in Polymers

The subtle differences in the microstructure of polymers which are manifested in the form of isomerism also affect the behaviour of a polymer. The isomerism in polymers can be grouped under four types:

1. Architectural
2. Orientational
3. Configurational
4. Geometric

Architectural

Differences are due to branching and network formation when polymer is synthesized from monomer having same molecular formula (i.e. isomeric monomers), e.g. the different CRU having same molecular formula C_2H_4O are

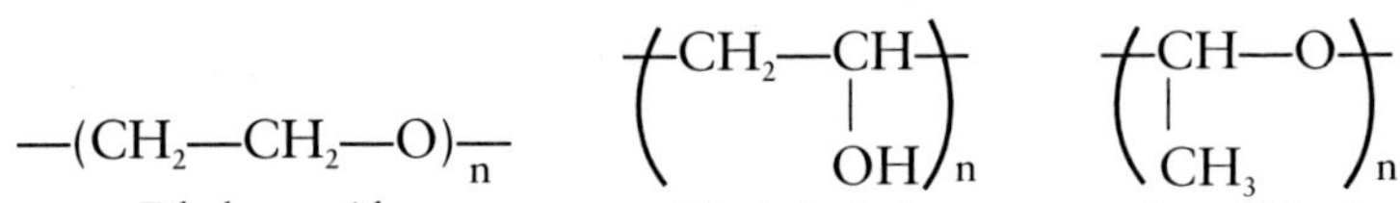

Ethylene oxide — Vinyl alcohol — Acetaldehyde

(a) (b) (c)

Fig. 1.2

These monomers generate polyethylene oxide, polyvinyl alcohol and polyacetaldehyde, respectively. All of them have different atomic arrangements as well as different physical properties, e.g. glass transition temperature of these polymers is 206 K, 358 K and 243 K, respectively.

Orientational

Whenever an asymmetric vinyl monomer is attached by a free radical, two possibilities arise:

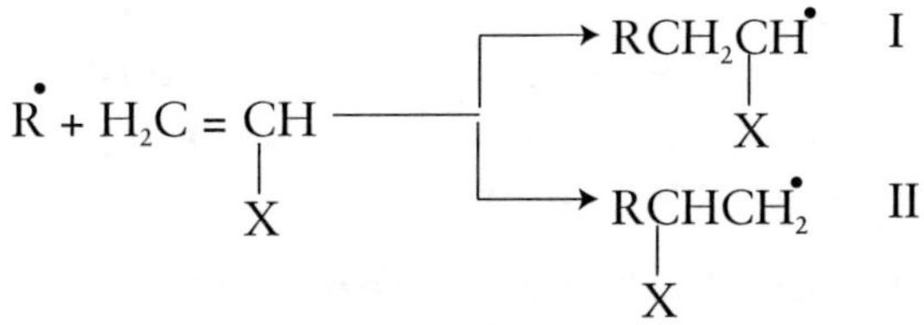

Fig. 1.3 (a)

The head of the free radical is marked by the carbon atom bearing the group X. All the three linkages, viz. head-to-head, head-to-tail and tail-to-tail can appear in a single molecule. But mostly in vinyl monomers, head-to-tail addition is preferred. The route I leads to a polymer having head-to-tail configuration since the successive addition of III repeat unit takes place in Fig. 1.3 (b).

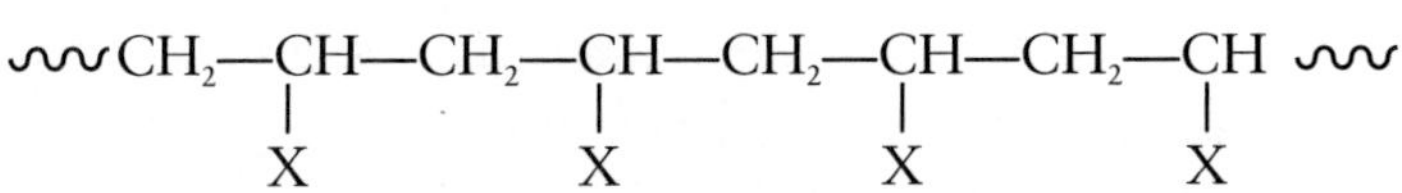

III

Fig. 1.3 (b)

While route II leads to head-to-head (or tail-to-tail configuration.

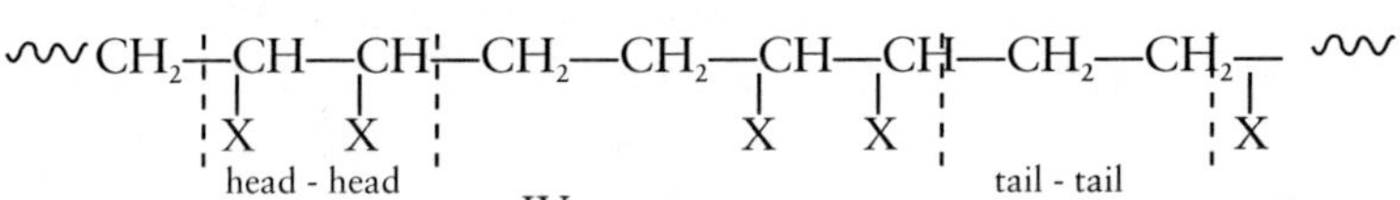

IV

Fig. 1.3 (c)

This preference to monomers to add on the free radical I or II depends on the stability of resultant X-mer, energy considerations and the steric hindrance to the approach of incoming CRU by the already present group X. The structure I has the possibility of resonance stabilization since X group is α to carbon atom having an unpaired electron. The hindrance caused by X is less. Such resonance stabilization lacks in structure II and the hindrance is more in head-to-head addition. However, route II is followed when termination is by the combination.

The stereo regular polymers show two types of isomerism—optical isomerism and geometrical isomerism. This stereoregularity arises due to the particular regularity of the main chain atoms or the substituent.

Optical Isomerism (Configurational Isomerism)

This isomerism develops in a molecule if it has asymmetric carbon atom. Asymmetric carbon atom is that which has all the four substituent different, e.g. substituted polyethylene.

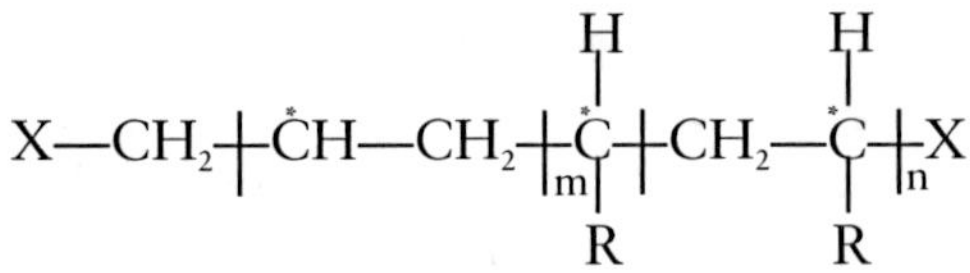

Fig. 1.4

In this molecule, every alternate carbon atom is asymmetric and is designated by *. This carbon atom has —H, —R and two polymer chain segments of chain length m and n. X denotes end groups.

This has a planar zig-zag carbon chain due to the bond angles with —H and —R groups above and below the plane of the paper. The —H and —R groups are interchangeable since the free rotation abound C—C single bond is allowed. Depending upon the position of R group (whether it is below or above the plane of the paper) the C* has a possibility of showing optical isomerism. The d- or l- form give rise to three different types of isomeric structures.

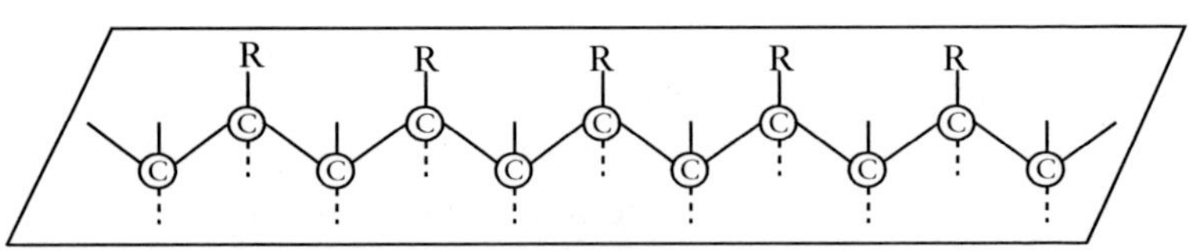

I (Isotactic) -d-d-d-d-d- or -l-l-l-l-l-

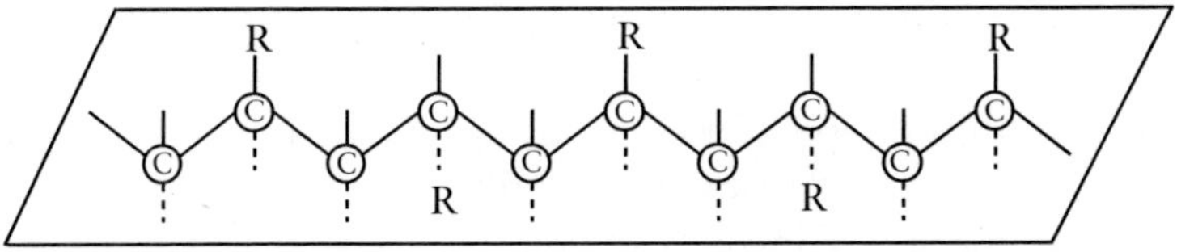

II (Syndiotactic) -d-l-d-l-d- or -l-d-l-d-l-

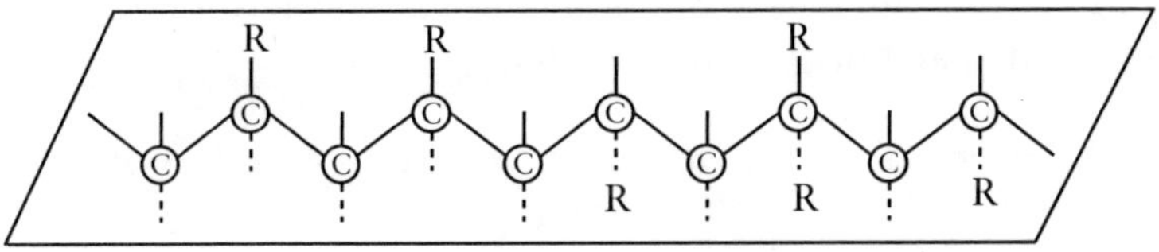

III (Atactic) -d-d-l-d-l- or -l-l-d-l-d-

Fig. 1.5

In structure I of Fig. 1.5, all the R groups are on one side of the carbon chain. In structure II of Fig. 1.5, R groups are alternately arranged above and below the carbon chain while in structure III of Fig. 1.5, the R groups are on randomly located. All the three structures have same chemical structures but are different in some of the physical properties.

When an asymmetric vinyl monomer of the type $CH_2 = CXY$ is polymerized where X and Y are two different substituent groups there exist two different stereoisomers of the repeat units.

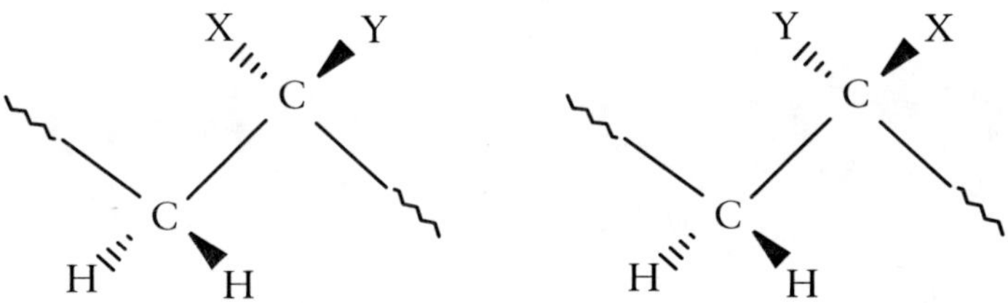

(a) (b)

Fig. 1.6

Carbon atom 2 is asymmetric of a chiral carbon atom as it is attached to four different groups. But both of these stereoisomers are not optically active because the structure of C atom 1 in both the cases is identical. However, the incoming repeat unit can have various orientations giving rise to different tacticity to the polymer.

1. If all the repeat units have the same configuration then they form an Isotactic polymer.
2. If the alternate repeat unit has altered structure then syndiotactic polymer is formed.
3. If random placement of incoming repeat unit takes place then atactic polymer results.

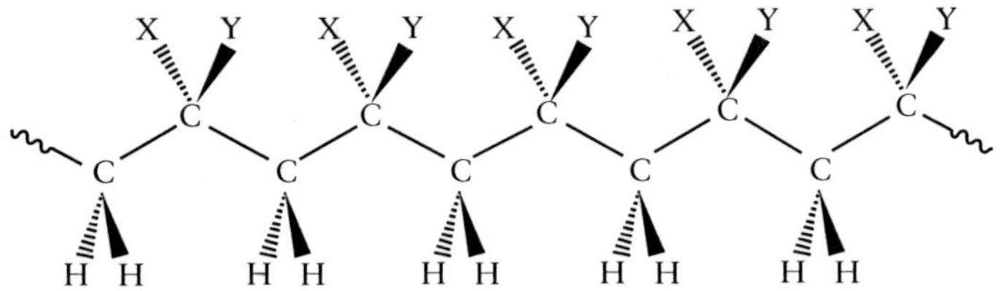

Isotactic

(a)

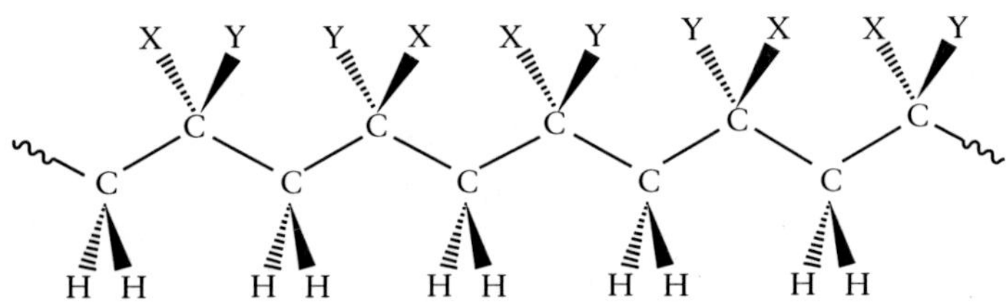

Syndiotactic

(b)

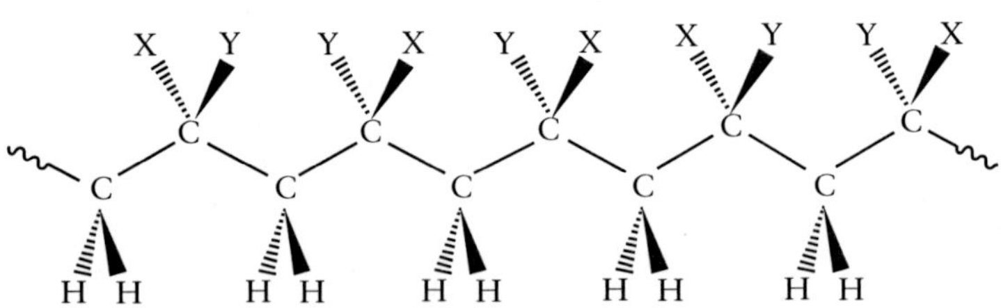

Atactic

(c)

Fig. 1.7

Things become more complicated when 1,2-disubstituted ethylenes (CHR = CHR) are polymerized. In this case, every carbon atom of the polymer chain becomes chiral. Among the polymer diatactic structures, two are isotactic and one is disyndiotactic. Among isotactic, one is erythro having all carbon atoms with same configuration and the other is threo with an alternate configuration. The formation of erythro and threo diatactic structures depends on the monomer taken cis-substituted monomer forms threo form and trans-substituted monomer forms erythro structure.

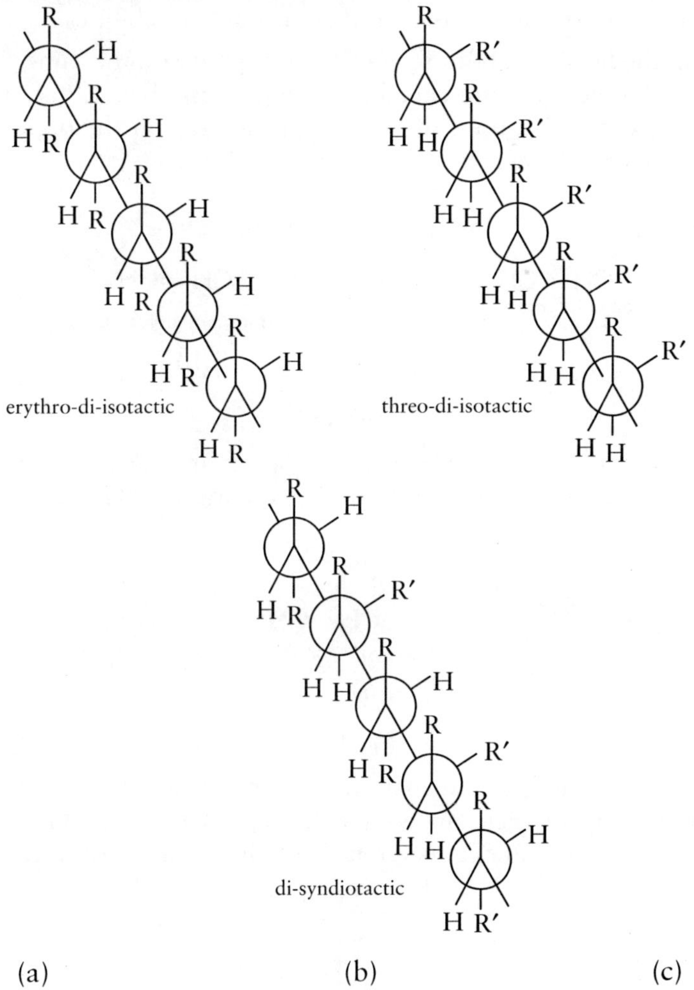

Fig. 1.8

Geometric Isomerism

Geometric isomerism develops due to the difference in the structure around a double bond. For example 1,3-butadiene. It has two double bonds but during polymerization, the repeat unit is left with one double bond.

Reaction

$$n\mathrm{CH_2{=}CH{-}CH{=}CH_2} \rightarrow \mathrm{+\!\!\!\!-CH_2{-}CH{=}CH{-}CH_2\!-\!\!\!\!+_{\,n}}$$

The repeat unit has two configurational possibilities.

In the first case, both the —CH_2 groups are on the same side of the double bond are called cis-configuration Fig. 1.9(a), and in the second case both the —CH_2 groups are on the opposite sides are called trans-configuration Fig 1.9(b).

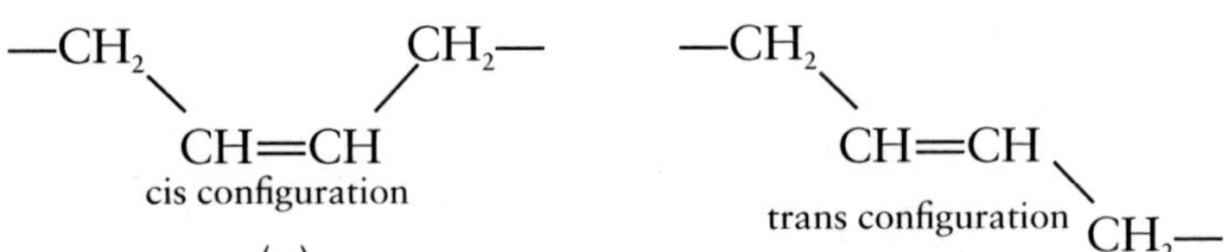

cis configuration (a)

trans configuration (b)

Fig 1.9

In the case of butadiene units, one more configuration is possible that is known as 1,2-vinyl configuration. (Fig 1.9(c)).

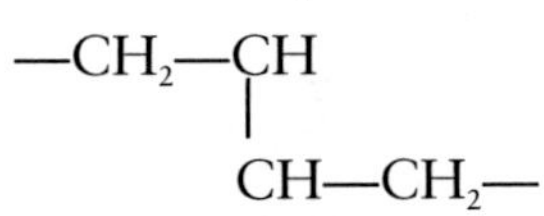

1, 2-vinyl configuration

(c)

Fig 1.9

When conjugated dienes are polymerized polymers with geometric isomerism are obtained, e.g. CH_2=CX-CH=CH_2; 2-methyl-1,3-butadiene (isoprene) polymerize in three different ways.

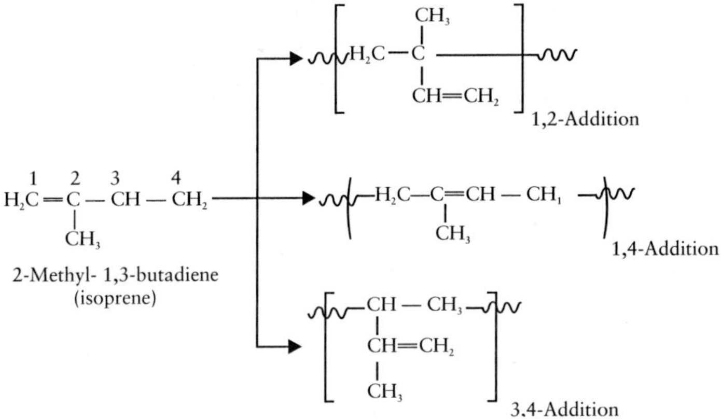

Fig. 1.9(a)

Addition can be via 1,2- or 3,4-mechanism. These result in isotactic, syndiotactic or atactic structures. But it is only 1,4 route which leads to cis-trans form

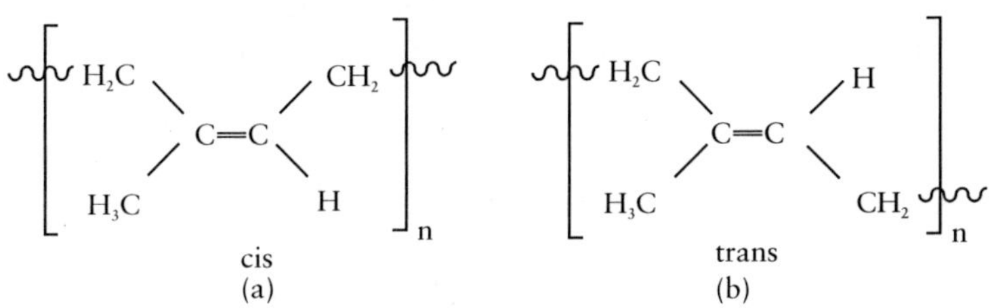

Fig 1.10

Widely used 1,3-dienes to synthesize polymers are

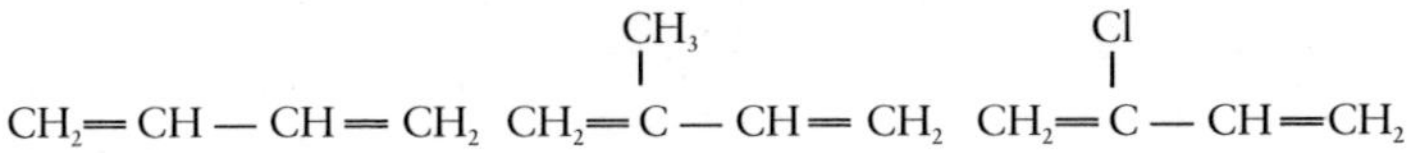

Butadiene Isoprene Chloroprene

They all have the general formula:

$$CH_2{=}CR{-}CH{=}CH_2$$

If in a polymer all the cis-configurations are found then it becomes like a spring and can be elongated. If all are trans, the polymer has straight and rod-like structure. However, in actual practice, mixture of cis and trans is obtained.

1.7 Rise of Polymer (Macromolecular) Science

History: Man's unending quest for new and improved materials has expanded with time and resulted in the discovery of polymers.

History of Development of Polymers

Period	Event	Polymers synthesised
1840 – 1920	Vulcanisation of natural rubber	Cellulose nitrate, Bakelite, phenol-formaldehyde
1920	Concept of macro-molecules	Alkyd resins, U-F and Thiokol resins
1926		Alkyds
1927		Plasticised PVC
1929		Urea – formaldehyde
1930		Melamine resins, Neoprene, Polyethylene, styrene-butadiene rubber
1936		Poly(vinylchloride), polychloroprene (neoprene), poly(methyl methacrylate), poly(vinyl acetate)
1937		Polystyrene
1939		66 – Nylon
1940		patented: PET
1941		Polytetrafluoroethylene
1942		Unsaturated polyesters (terylene, Teflon)
1943		Branched polyethylene, butyl rubber, 6 - nylon, silicones
1944		Poly(ethylene terephthalate)
1947		Epoxies
1948		ABS resins
1950		Spandex (polyurethane elastomers), HDPE, stereoregular PP, polycarbonates
1953	Nobel Prize — Hermann Staudinger	

Period	Event	Polymers synthesised
1955		Linear polyethylene
1956		polyoxymethylene
1957		Polypropylene, polycarbonate
1960		Aramides, flame retardant polymers and carbon fibres, thermoplastic elastomers
1963	Nobel Prize — Giulio Natta, Karl Ziegler	
1964		Ionomers resins
1965		polyimides
1970		Thermotropic liquid crystalline polymers, thermoplastic elastomers
1972		Acrylonitrile copolymers, mouldable polyesters
1974	Nobel Prize — Paul J. Flory	Aromatic polyamides
1980		Polyetheretherketone (PEEK)
1984	Nobel Prize — Bruce Merrifield	
1990		Polymer composites, high-performance polymers, biodegradable polymers, water soluble polymers, conducting polymers, electroluminescent polymers, multifunctional star and dendrimers
1991	Nobel Prize — Pierre-Gilles de Gennes	
2000	Nobel Prize — Hideki Shirakawa, Alan G. Mac Diarmid, Alan J. Heeger	
2002	Nobel Prize — J.B. Fenn, K. Tanaka, K.T. Wuthrich	

The most widely used commercial polymers are handled by different type of people worldwide hence pictorial symbols have been assigned to them by The Plastic Bottle Information Bureau,

Washington, DC. These symbols are put on the containers for the convenience of the user.

Code	Letter I.D.	Polymer Name
♳	PETE	Poly(ethylene terephthalate)
♴	HDPE	High-density polyethylene
♵	V	Poly(vinyl chloride)
♶	LDPE	Low-density polyethylene
♷	PP	Polypropylene
♸	PS	Polystyrene
♹	Other	Different polymers

1.8 Binding Forces in Polymers, Molecular Forces and Chemical Bonding in Polymers

The structural aspects of a polymer can be understood at three different levels.

1. The primary structure (chemical structure of the monomer or CRU)
2. The secondary structure (single polymer chain) and
3. The tertiary structure (aggregation of polymer chains).

Physical properties of polymers are affected by the size of the molecules and by the nature of the primary and secondary forces. The amount of symmetry and molecular structure also influence the characteristics of the polymer. Secondary forces become important in the case of polymers owing to their large size.

1. **The primary structure,** i.e. the structure of basic unit, viz. a monomer and more precisely the CRU, is made up of atoms that are tied together by strong bonds to form small molecules using valence electrons. There are three types of primary bonds: ionic, metallic, and covalent. The atoms of a polymer are mostly bonded together by covalent bonds that are formed between atoms of similar electronegativity. The bond length of primary bonds is between 0.1-0.22 nm (carbon—carbon bond length 0.15-0.16 nm) and the bond energy between 320-370 kJ per mol (for single bond). In order to know about the properties and structure of polymer, one should know about the chemical

nature of monomers, the reactivity of monomers and the mode of polymerization.

2. **The secondary structures** occur because of the length of polymer chains and there exist intramolecular forces between same chains and/or intermolecular forces with the neighbouring chains, where the distances are 3-4 Å. These intramolecular attractions do not involve valence electrons and hence are weak and are termed as secondary bonds. Heating or dissolving in a liquid can break these bonds. The strength of secondary bonds is influenced by the functional groups present in polymers. If non-metals like nitrogen, oxygen, fluorine and chlorine are present in the side chain, they shift the shared electrons away from carbon and hydrogen, resulting in the polarization of the molecule. The positive poles of one polymer chain attract the negative poles of the other polymer chain.

Secondary forces are of several types:

(a) Dipole forces

(b) Hydrogen bonds

(c) Induction forces

(d) Van der Waals (dispersion) forces

Generally, hydrogen bonding and dipole forces are way stronger than van der Waals forces, but it is van der Waals forces that play a major role in binding the two polymer chains together.

Dipole Interactions (Polar Interactions): If in a molecule a covalent bond is formed between an electronegative atom like nitrogen, oxygen, fluorine and chlorine, and carbon (or hydrogen), electrons of a covalent bond shift away from carbon (and hydrogen). As a consequence, partial negative and positive charges develop, i.e. the molecule becomes partially polarized. There develops a force of attraction between the positive and the negative centres of neighbouring polymer chains. Positive and negative charges remain on the covalent bond and can be felt as static electricity if polymer fibers are rubbed together. Dipole interactions are strongly influenced by temperature.

Hydrogen Bonds: A hydrogen bond is a specific type of dipole interaction. This bond is formed between hydrogen (which is temporarily becomes positively charged) and small-sized electronegative atoms like F, O, N, etc. In the case of polymers hydrogen bonding takes place between functional groups like —CO, —OH, —NH_2, —NH, —ethers, etc. hydrogen bonding is important in biopolymers, e.g. proteins.

Molecules with either —N—H or —O—H groups have comparatively strong secondary bonds because of their ability to form hydrogen bonds.

Induction Forces (Non-polar Interactions): As valence electrons move around the nuclei, sometimes the electron cloud of non-polar polymer molecule becomes temporarily imbalanced and the positive and negative parts are formed in the molecule. This makes the molecule polar, which has a force of attraction. However, these forces of attraction are weak. The ease with which the +ve and –ve displacement takes place, it is known as polarizability of the molecule, which is independent of the temperature.

Van der Waals Forces: It is possible in non-polar molecules that during rotation by electrons around nucleus the centre of electron cloud may not coincide with the centre of nuclei at a particular instant of time. This causes an instantaneous dipole in the molecule which is a time-dependent dipole moment. If this dipole of one molecule gets aligned with the instantaneous dipole of another molecule and an intermolecular attraction develops between both the molecules called dispersion forces. This force is also known as van der Waals force.

The Tertiary Structure: This refers to the three-dimensional structure of a polymer. The polymer molecules aggregate to form either a crystalline solid or an amorphous mass. It is the secondary intermolecular forces which bind the molecules together. Though the secondary forces are very week but because of the large size of the polymer molecules, they become very strong. In addition, the chain entanglement also plays a part in further strengthening the binding of polymer molecules.

1.9 Chemical and Physical Properties of Polymers

1.10 Molecular Weight, Average Molecular Weight, Molecular Weight Distribution and Degree of Polymerization

Polymer molecule has high molecular weight, which is represented by 'M'. The molecular weight of polymers lies in the range of 10^3-10^7 or even greater. The units are g per mol or amu per molecule.

All the molecules of low molecular weight simple compound have identical molecular weight. Such compounds are called monodisperse. But this is not the case with polymers. A polymer is made up of molecules that have a range of molecular weight values, which are termed as polydisperse.

Degree of Polymerization: The degree of polymerization of a polymer is the number of repeat units in one molecule. It is represented by 'n', which is the ratio of the molecular weight of the polymer to the molecular weight of the repeat unit: $n = M/M_0$, where M = molecular weight of polymer and M_0 = the molecular weight of repeat unit.

In addition polymers, the successive repeat units are add-on to the chain, no side product is formed, so the weight of the monomer and that of the CRU are identical and the molecular weight of the polymer is a simple multiple of the molecular weight of monomer.

Hence, molecular weight of polymer $M = M_0.n$, where M = the molecular weight of the polymer, M_0 = the molecular weight of monomer and n = the number of CRU.

In the condensation polymers, one or two small molecules are eliminated for each chain linkage formed. Elimination of molecules like water, HCl, etc. takes place. Hence, the molecular weight of the monomer and the repeat unit, are different. For example, a dibasic acid (with subscript A) reacts with a diol (with subscript B) to produce a polyester, and one water molecule is liberated during every —COOH and —OH combination. The molecular weight of the polyester will differ from the combined weight of the reactants by twice the molecular weight of the water. Therefore:

$$n = \frac{M}{M_0} = \frac{M}{M_A + M_B - 2M_{H_2O}} \quad \text{...(1.1)}$$

Molecular Weight Distribution: Synthetic polymer samples contain chains of various chain lengths and hence have a variety of molecular weights. If a graph of molecular weight of polymer chains of a particular sample and number of molecules that have a particular molecular weight is plotted, the trend of the molecular weight distribution of that sample will be obtained. This distribution has been depicted in the following figure:

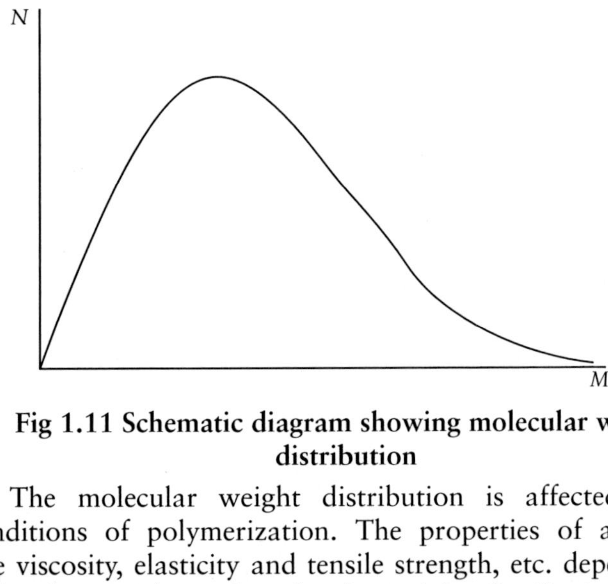

Fig 1.11 Schematic diagram showing molecular weight distribution

The molecular weight distribution is affected by the conditions of polymerization. The properties of a polymer like viscosity, elasticity and tensile strength, etc. depend on its molecular weight and molecular-weight distribution. Hence, the knowledge of molecular weight and molecular weight distribution becomes an important parameter. But it is not sufficient to know only the average molecular weight. It is also important to mention here that polymers with the same structures can have very different molecular weight distributions and properties when made by different mechanisms.

Average Molecular Weight: During the formation of polymers, molecules have different degrees of polymerization, i.e. different chain lengths. Thus, molecules in a sample have different molecular masses. Due to this reason, molecular weight of a polymer is represented as the average of all these molecular weights. The general equation for average molecular weight $\bar{M}$ is

$$\bar{M} = \frac{\Sigma_i N_i M_i^{\alpha}}{\Sigma_i N_i M_i^{\alpha-1}} \qquad ...(1.2)$$

where N_i = the number of moles of molecules of molecular weight M_i, α = weighing factor that defines average of the molecular weight distribution.

The weight $W_i = N_i M_i$

The various representations of molecular weights that help in determining properties of the polymers are as follows. The method of their determination is also different.

1. Number-average molecular mass ($\bar{M}_n$)
2. Mass-average molecular mass ($\bar{M}_w$)
3. Z-average molecular mass ($\bar{M}_z$)
4. Viscosity-average molecular mass ($\bar{M}_v$)

Number-average Molecular Mass ($\bar{M}_n$): When the total molecular mass of all the molecules of a sample is divided by the total number of molecules, the result obtained is Number-average molecular mass ($\bar{M}_n$). Value of $\alpha = 1$.

Suppose in a particular sample N_1 molecules have molecular mass M_1, N_2 molecules have molecular mass M_2, N_3 molecules have molecular mass M_3 and so on. Then we have a total molecular mass of all the N_1 molecules = N_1M_1, a total molecular mass of all the N_2 molecules = N_2M_2, a total molecular mass of all the N_3 molecules = N_3M_3 and so on.

Hence, the total molecular mass of all the molecules = $N_1M_1 + N_2M_2 + N_3M_3 + \text{---} = \Sigma N_iM_i$ ...(1.3)

The total number of all the molecules = $N_1 + N_2 + N_3 + \text{---} = \Sigma N_i$ ...(1.4)

On dividing equation (1.3) by (1.4)

$$\bar{M}_n = \frac{\Sigma N_i M_i}{\Sigma N_i} \qquad ...(1.5)$$

Mass-average molecular mass $(\bar{M}_w)$: When the total molecular masses of groups of molecules having particular masses are multiplied with their respective molecular masses, the products are added and the sum is divided by the total mass of all the molecules the result obtained is called the mass-average molecular mass $(\bar{M}_w)$. Earlier the term 'weight average molecular weight' was used which was later replaced by 'mass average molecular mass' represented by $(\bar{M}_w)$. Molar masses can also be used instead of molecular mass. Value of $\alpha = 2$.

Condition under which $\bar{M}_n = \bar{M}_w$

We know

$$\bar{M}_n = \frac{N_1 M_1 + N_2 M_2 + ----}{N_1 + N_2 \pm ----} \qquad ...(1.6)$$

and $$\bar{M}_w = \frac{N_1 M_1^2 + N_2 M_2^2 + ----}{N_1 M_1 + N_2 M_2 + ----} \qquad ...(1.7)$$

If $\bar{M}_n$ is to be equal to $\bar{M}_w$ then

$$\frac{N_1 M_1 + N_2 M_2 + ----}{N_1 + N_2 \pm ----} = \frac{N_1 M_1^2 + N_2 M_2^2 + ----}{N_1 M_1 + N_2 M_2 + ----} \qquad ...(1.8)$$

The two sides are equal only when $M_1 = M_2$.

That is all the molecules have the same molecular mass. But in that case, the average has no significance.

It can be observed that number-average molecular mass is simply the arithmetic mean of all the molar masses whereas the mass-average molecular mass is the sum of the fractional masses that each molecule in a given sample contributes to the average according to the ratio of its mass to that of the whole sample. In the number-average molecular mass each molecule counts equally but in the mass-average molecular mass, molecules contribute according to their masses.

Z-average Molecular Mass $(\bar{M}_w)$: The equation for Z-average Molecular Mass is

$$\bar{M}_z = \frac{\Sigma N_{1i} M_i^3}{\Sigma N_1 M_i^2} \quad ...(1.9)$$

Value of $\alpha = 3$.

Since the weighing factors in the Z-average molecular mass is $\Sigma N_{1i} M_i^2$, the molecules having higher molecular mass are weighed even more heavily resulting $\bar{M}_z > \bar{M}_w > \bar{M}_n$.

Viscosity-average Molecular Mass $(\bar{M}_v)$: It is defined as

$$\bar{M}_v = \left[\frac{\Sigma_i N_i M_i^{a-1}}{\Sigma_i N_i M_i}\right]^{\frac{1}{a}} \quad ...(1.10)$$

The constant 'a' is known as Mark-Houwink exponent. It is characterized by the system under investigation and generally lies in the range 0.5 < a < 1.0. Note that when a = 1, $\bar{M}_v = \bar{M}_w$. The viscosity-average molecular mass is not an absolute value, but a relative molar mass based on prior calibration with known molar mass for the same polymer-solvent-temperature conditions. The variable 'a' depends on all these three conditions.

Polydispersity Index: A polymer sample has molecular chains of different lengths, this heterogeneity is called polydispersity. The mass-average molar mass $(\bar{M}_n)$ is always larger than the number-average molar mass $(\bar{M}_n)$ unless the sample is monodisperse in which case $\bar{M}_n = \bar{M}_w$. The ratio is called the polydispersity index PDI of a polymer sample.

1.11 Macromolecules vs Polymers

Macromolecules and polymers are giant molecules. Macromolecules comprise both polymeric (made up of small repeating units called monomers) and non-polymeric molecules (i.e. they cannot be divided into smaller units). Lipids are an example of macromolecules that aren't polymers, as their structure does not consist of a repeating chain of monomers. Proteins, carbohydrates and nucleic acids are all macromolecules. Colloids, macromolecules and micelles lie between the categories of ordinary particles and ordinary molecules.

Part – B

Polymerization

2 Chain Growth Polymerization (Addition Polymerization)

2.1 Synthesis of Polymers

As we have already seen that, polymers usually contain carbon backbone. These carbon atoms are linked together by covalent bonds (sometimes the backbone also contains elements other than carbon, as has been discussed earlier in the classification). The side chain along with hydrogen contains other atoms such as oxygen, nitrogen, halogens, etc. This gives polymers a wide variety. Hence, the polymerization of monomers to form polymers should be studied systematically. W.H. Carothers in 1929, proposed two major types of polymers based on polymerization methods, which are used to convert small molecules (monomers) into polymers. 1. Addition Polymers and 2. Condensation Polymers.

1. *Addition Polymers*—they are synthesized by addition polymerization in which no liberation of any small molecule occurs.

2. *Condensation Polymers*—they are synthesized by Condensation polymerization and involve elimination of small molecules like water, HCl, NH_3 etc. during each step of propagation.

With the development of polymer science, new polymers were discovered. During their synthesis, a number of exceptions arose, which forced to lose the term 'condensation polymerization'. The new term introduced was 'step growth polymerization' or simply 'step polymerization'. The term step polymerization included the polymerization of polymers like polyurethanes, which are formed without the elimination of a small molecule.

And the term addition polymerization is replaced by the new term 'chain growth polymerization' or 'chain polymerization'. Addition polymerization follows the same mechanism as that of free radical chain reactions.

This new terminology is based on the mechanism of polymerization in addition to the stoichiometry of the polymerization and composition of the main chain of the polymer.

2.1.1 Introduction of Addition Polymerization

Usually, unsaturated monomers are the starting material for synthesizing addition polymers, which takes place through chain reaction; hence, it is named chain polymerization. In chain polymerization, an initiator is used, which requires a small amount of energy to form a reactive centre. This initiator reactive centre reacts with the monomer in such a way that the reactive centre is maintained at the end of the growing polymer chain and can continue to react with added number of molecules of the monomers at a faster rate. This leads to the formation of high molecular weight polymer while monomer molecules still remain in the reaction mixture. One can soon find high molecular weight polymer and unchanged monomer in the reaction mixture. But in the reaction mixture, there is no polymer with intermediate molecular weights. This reaction is little affected by the extent of reaction.

The mechanism of addition polymerization can be of two types: one in which there active centre needed for the initiation of the reaction is a free radical and the other in which an ionic centre is formed. There are several other types of chain polymerizations too. One of the most important chain polymerizations is where the active centre is formed on a catalyst that might involve metal ions that act as a coordination catalyst rather than an initiator. The other worth mentioning is chain-opening polymerization.

2.2 Mechanism of Addition Polymerization

In chain-growth polymerization, like other chain reactions have following three steps:

1. Initiation
2. Propagation
3. Termination

In initiation step, active centre in the monomer is generated by an initiator; in propagation, addition of monomer to the growing chain at a fast rate takes place without the loss of active center and in termination, the active centre is destructed. In some cases, there can be chain transfer reaction also, where the active centre is transferred to another species present in the reaction mixture.

Initiation: Polymerization takes place in the presence of the compounds called initiators. An initiator on dissociation generates active centres, which add on the monomer molecules rapidly. Active centres can either be free radical or non-radical. Non-radical active centre includes ionic (cationic or anionic) active centre and coordination active centre, depending on the kind of initiator system that has been used. Based on the nature of the initiator used, the chain-growth polymerization is classified as follows:

1. Free radical polymerization

   Among non-free radical polymerization we have—
2. Ionic polymerization—it can be of two types:
   (a) Cationic polymerization
   (b) Anionic polymerization
3. Coordination or stereoregular polymerization and
4. Ring-opening metathesis polymerization

The reactivity of a monomer towards the initiator depends on the polarity of the monomer and the Lewis acid-base strength of the ion formed.

The presence of conjugation or moderately electron withdrawing group on the monomer favours free radical polymerization.

The ionic mechanism involves an attack on the π electron pair of the monomer by positive or negative ions. In cationic as well as anionic polymerizations, the gegen ion remains in the proximity of the growing centre, hence, their initiation and propagation steps show similar characteristics.

There are monomers that can follow more than one route for polymerization. Choice of the route depends on the following factors:

(a) Polarity of the monomer and

(b) The Lewis acid-base strength of the ion formed.

If the double bond of the monomer has an electron-donating group, it will form a stable carbonium ion, i.e. it acts as a Lewis base (have negative charge) hence can be catalyzed by cationic catalysts and favours cationic polymerization, e.g. Friedel-Crafts catalysts like aluminium chloride $AlCl_3$, BF_3. The monomers that are polymerized with these catalysts are styrene $C_6H_5CH\text{-}CH_2$, isobutylene, etc.

If an electron withdrawing group is attached on the monomer, it forms a stable carbanion (have +ve charge) and polymerization takes place with anionic catalysts and favours anionic polymerization, e.g. acrylonitrile.

If the monomer can be initiated by a coordination compound then coordination polymerization can be the chosen method.

Propagation step, the reaction between the polymeric growing centre and the monomer molecule is the distinguishing feature among the four classes of chain-reaction polymerization.

2.2.1 Free Radical Polymerization

A free radical is an entity that contains an odd number of electrons but does not carry an electric charge.

Initiation: Initiation consists of the formation of free radical active centre. There are many organic molecules containing unstable bonds which can be dissociated thermally into numbers of free radicals. The thermal decomposition of initiator is an easy method for the generation of radical but one cannot have immediate control over the decomposition rate. On the other hand, photo-decomposition can be carried out with precision as the intensity of light is the controlling factor. High energy radiations are also used to carrying out polymerization.

This takes place in two steps—first is the formation of free radicals from an initiator and the second is the addition of this free radical to a monomer molecule. The free radicals can be formed by two methods: 1. Homolysis of the single bond producing two free radicals, and 2. Is the transfer of a single electron to or from an ion of the molecule (single –electron transfer processes), e.g. in redox reactions. This method sometimes produces single free radical species.

Homolysis is usually carried by thermolysis. The compounds containing peroxide (—O—O—) and azo (—N ≡ N—) linkages undergo homolysis to produce free radicals.

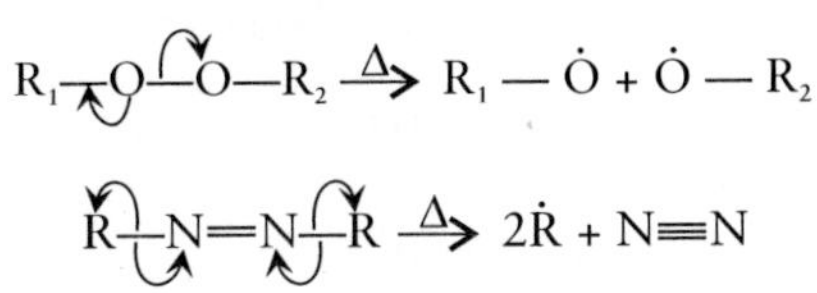

where R_1 and R_2 are aromatic, alkyl or H, R_1 can be = R_2, R is aromatic or alkyl.

Some of the thermal initiators are

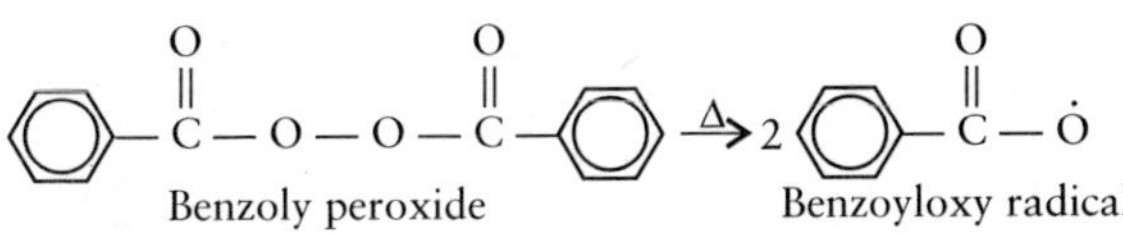

$$(CH_3)_3\,C - O - OH \xrightarrow{\Delta} (CH_3)_3\,C - O - \dot{O} + \dot{O}H$$

t-Butyl hydroperoxide — *t*-Butoxy radical — Hydroxyl radical

Sometimes the radicals produced undergo further breaking

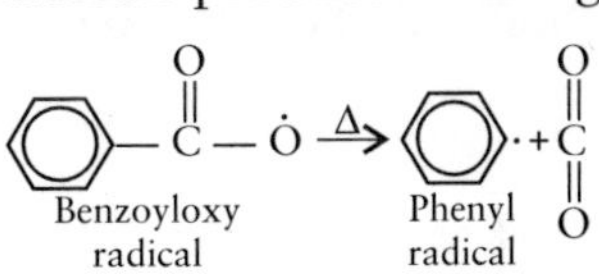

Homolysis can be brought about photolytically. Azobisisobutyronitrile dissociation can also take place by photolysis the mechanism is same as that of pyrolysis.

$$(CH_3)_2\underset{\displaystyle CN}{\underset{|}{C}} - N{=}N - \underset{\displaystyle CN}{\underset{|}{C}} - (CH_3)_2 \xrightarrow{h\nu} 2(CH_3)_2\underset{\displaystyle CN}{\underset{|}{\dot{C}}} + N{\equiv}N$$

Azobisisobutyronitrile — 2-Cynopropyl radical

If polymerization is to be carried out at low temperatures (below 50°C) then redox initiators would be preferred. The most popular redox initiators are cumyl hydro peroxide/Fe^{2+} system iron gives an electron to —H of the other species and O—O bond breaks homolytically. The OH fraction gains an electron for form hydroxyl ion. The other electron of the O—O bond goes to other oxygen atom forming cumyl oxy radical.

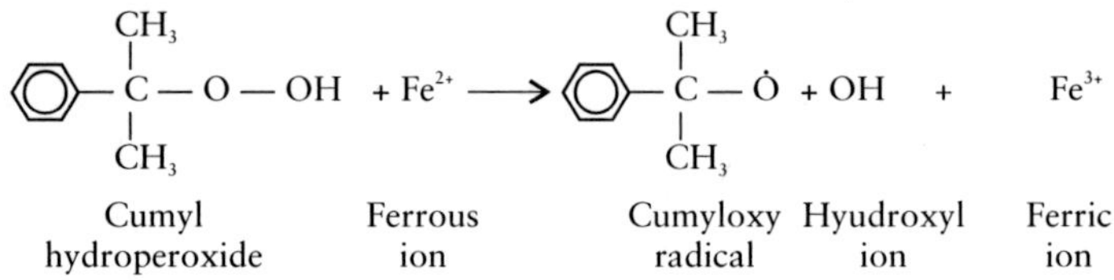

All other peroxides and hydroperoxides produce R—O radical, alkoxide and Fe^{3+} in a similar manner. Fe^{2+} can be replaced by other transition metals.

Per-sulphate and bisulphate ion initiator are a purely inorganic initiator system. Its reaction is as follows:

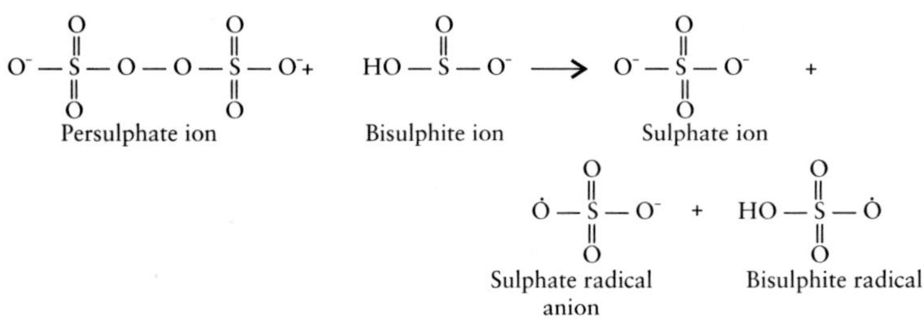

Once the active centre of initiator is created represented by $\dot{R}$, it attacks the π –bond of the monomer molecule. There are two possibilities

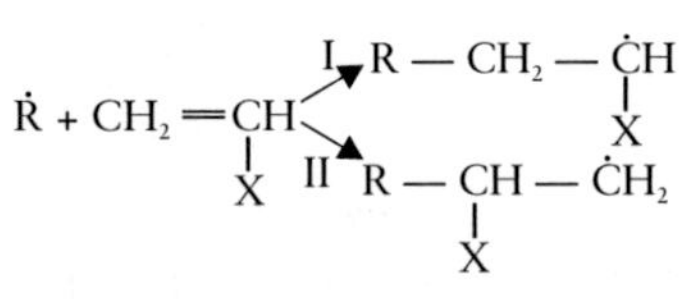

The route I is preferred as carbon of —CH_2 group is sterically less hindered. Group X also provides mesomeric stabilization to this radical, which makes the product I to be more stable.

Sometimes the initiator radical is not fortunate enough to react with the monomer. It is lost in a side reaction known as induced decomposition. This is a waste of initiator.

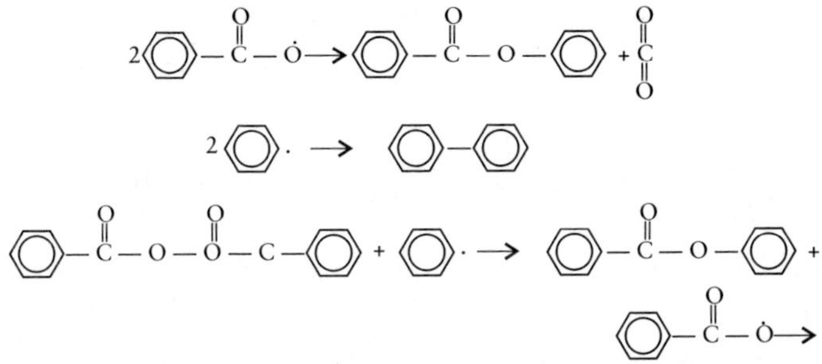

The general mechanism of free radical can be given as follows:

Mechanism

1. Initiation

$$CH_2{=}CH_2 + \dot{R} \rightarrow \dot{C}H_2{=}CH_2{-}R$$

2. Propagation

$$CH_2{=}CH_2 + CH_2{=}CH_2{-}R \rightarrow \dot{C}H_2{-}CH_2{-}CH_2{-}CH_2{-}R$$

$$CH_2{=}CH_2 + \dot{C}H_2{-}CH_2(CH_2)_2R \rightarrow \dot{C}H_2{-}CH_2(CH_2)_{n+2}R$$

3. Chain transfer

$$CH_2{=}CH_2 + \dot{C}H_2{-}CH_2(CH_2)_nR \rightarrow \dot{C}H_2{=}CH_3 + CH_2{=}CH(CH_2)_nR$$

4. Termination

(a) combination

$$R(CH_2)_nCH_2{-}\dot{C}H_2 + \dot{C}H_2{=}CH(CH_2)_nR$$
$$\downarrow$$
$$R(CH_2)_nCH_2{-}CH_2{-}CH_2{-}CH_2(CH_2)_nR$$

(b) disproportionation

$$R(CH_2)_nCH_2{-}\dot{C}H_2 + \dot{C}H_2{-}CH(CH_2)_nR$$
$$\downarrow$$
$$CH_2{=}CH(CH_2)_nR + CH_3{-}CH_2(CH_2)_nR$$

Let's take an example of polymerization of Styrene initiated by tertiary butyl peroxide.

First of all, the initiator, tertiary butyl peroxide decomposes on heating to generate two radicals per molecule.

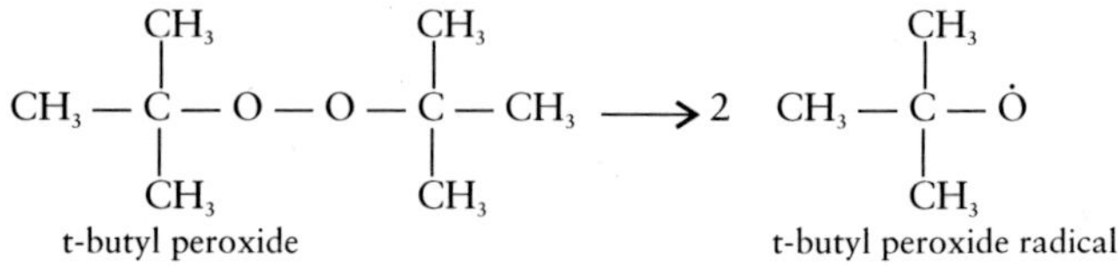

Since the initiator molecule, viz. tertiary butyl peroxide gets consumed; hence, it is called an initiator and not a catalyst.

1. Initiation: The radical initiator reacts with styrene molecule and another radical is formed.

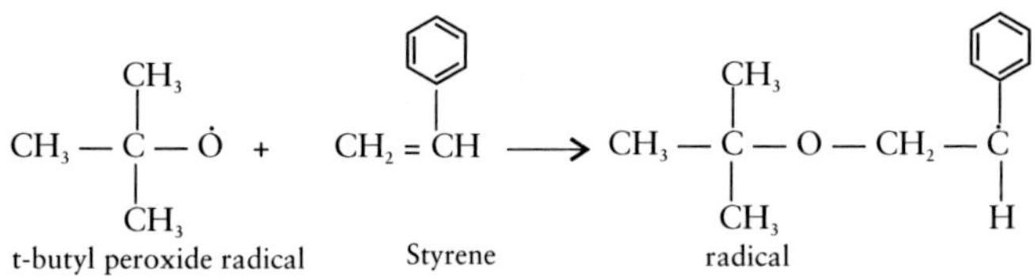

t-butyl peroxide radical Styrene radical

The unpaired electron can be placed either on the carbon containing the phenyl ring or on the methylene carbon of styrene. The unpaired electron will prefer to stay on the carbon atom that will form a more stable radical. Hence, the unpaired electron will be on the terminal carbon atom having phenyl ring as shown in the diagram.

2. Propagation: The next step is propagation, in which the free radical centre that is formed in the initiation step continues to add more styrene molecules to form a big polymer molecule having head-to-tail linkage and the phenyl rings occupy alternate carbon on the carbon atom chain (as a pendant).

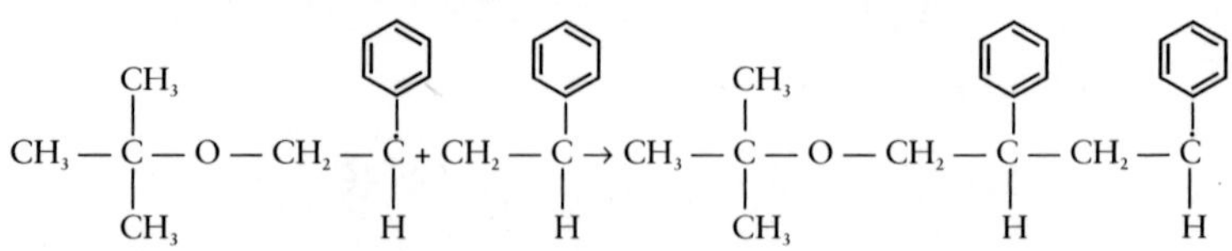

General propagation step

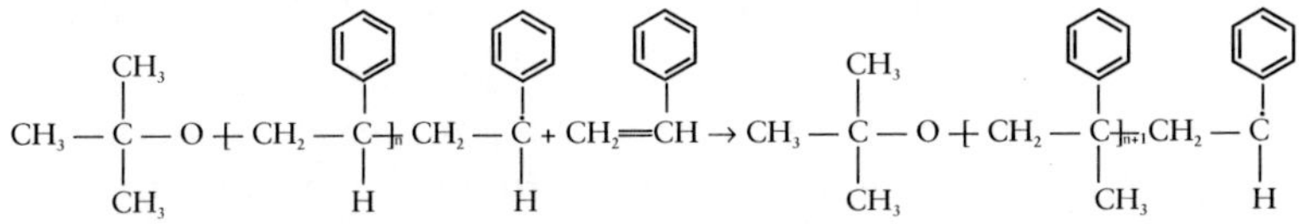

3. Termination: Propagation continues till the monomer lasts while the free radical centres could tend to react together and stop the propagation. This is known as termination. Termination takes place by two methods: one is combination or coupling and the other is disproportionation.

(a) ***Combination or coupling***: There are chances that the free radical centres of growing chains react together to form a single polymer molecule and thus polymerization stops resulting in a big polymer molecule.

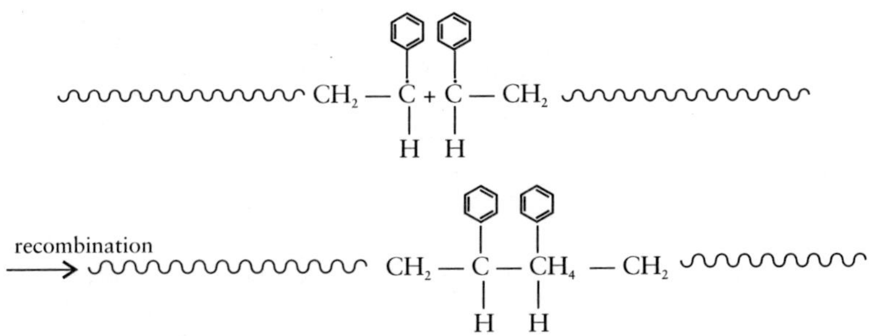

(b) ***Disproportionation***: However, hydrogen could be abstracted by another growing chain from a radical alpha and result in two stable non-reactive chains. One of which will have a saturated end while the other an unsaturated terminal olefinic end.

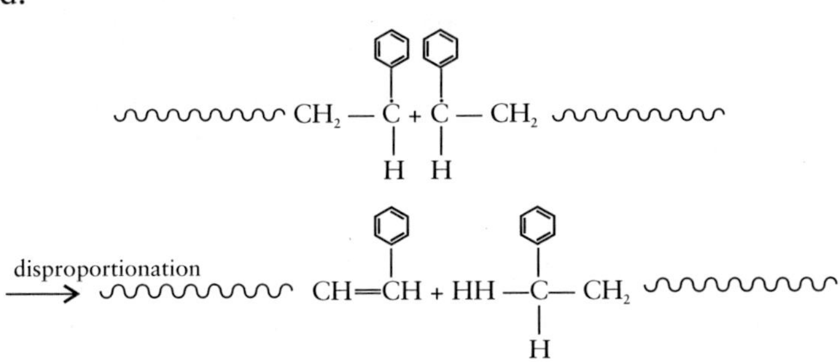

The two termination steps result in different molecular lengths and the kinetics of both will be different. These steps are temperature dependent.

Let us take another example for further illustration the reaction of styrene and benzoyl peroxide in a solution. On heating, benzoyl peroxide dissociates into two molecules of benzyloxy radicals which combine with the styrene monomer to give growth centres as follows:

Dissociation of initiator:

$$C_6H_5-\overset{O}{\overset{\|}{C}}-O-O-\overset{O}{\overset{\|}{C}}-C_6H_5 \xrightarrow{60\text{-}100°C} 2C_6H_5-\overset{O}{\overset{\|}{C}}-\dot{O}$$

Initiation:

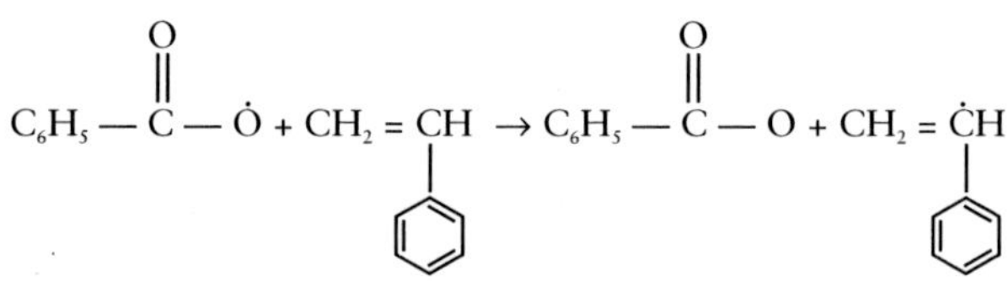

Propagation:

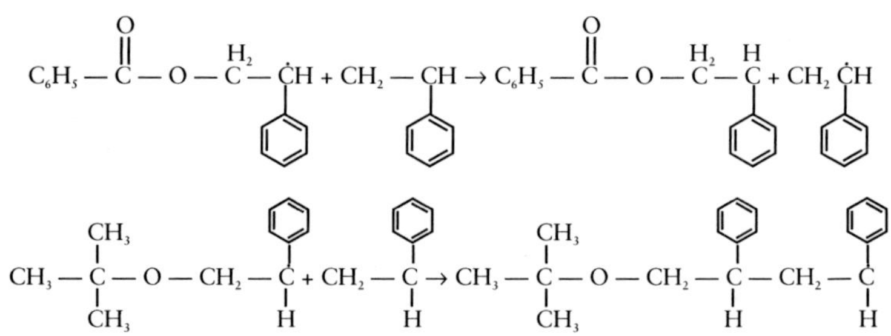

In general, the propagation reaction can be written as follows:

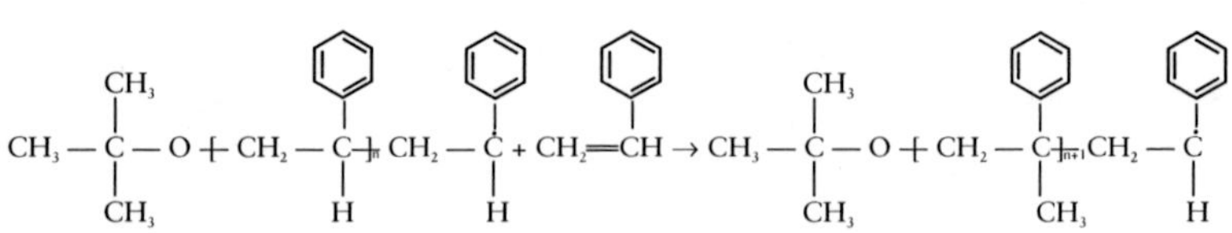

Chain transfer reaction: A reaction known as the chain transfer is also observed in some of the radical polymerizations.

Its active centre can be transferred either to a monomer, a polymer, a solvent or even to an impurity growth of the original chain stops. This new reactive species starts forming another polymer.

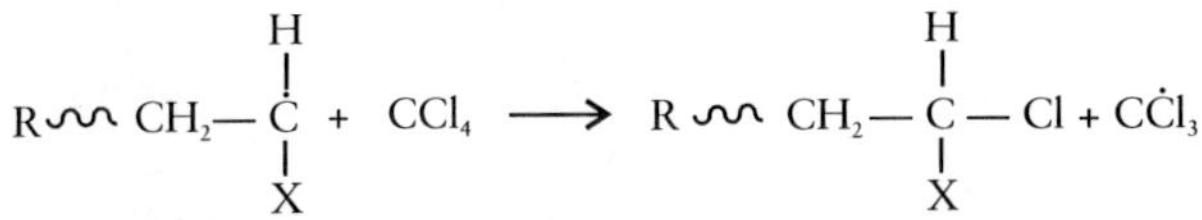

But if the molecule is unsaturated then either of the following two processes will take place.

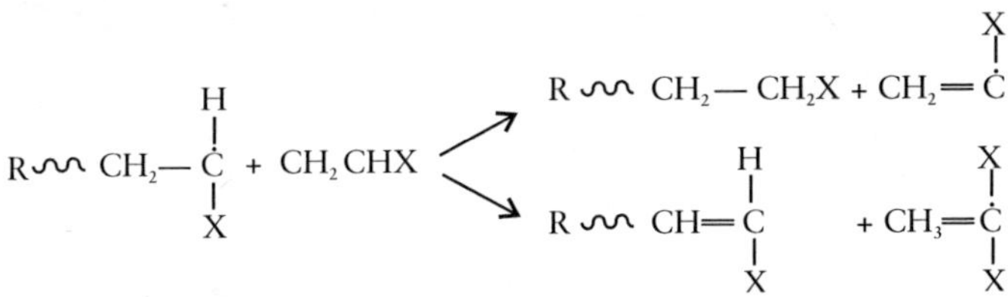

Inhibitors and Retarders: *Inhibitors*: They are the substances that can kill or inhibit the chain growth by combining with free radicals and forming either stable products or inactive free radicals. Normally, inhibitors react with initiating radicals to form new free radicals which are very stable. These are unable to initiate new chain. The reactivity of an inhibitor depends upon—

1. Chain transfer constant of the inhibitor for a particular polymerization reaction
2. Reactivity of radical formed by inhibitor
3. Reactivity of the monomer is taken.

Phenols, quinones, hydroquinones, nitrobenzene, dinitrobenzene, benzothiazine and aromatic amines are commonly used inhibitors.

Inhibitors are added to the monomer to prevent polymerization during the transportation. The inhibitor should be destroyed and removed before use. The monomer is either distilled or washed with aqueous solution of alkali. On a large scale, the inhibitor is destroyed by adding more quantities of initiators.

As an example the inhibitory reaction of nitrobenzene can be shown as follows:

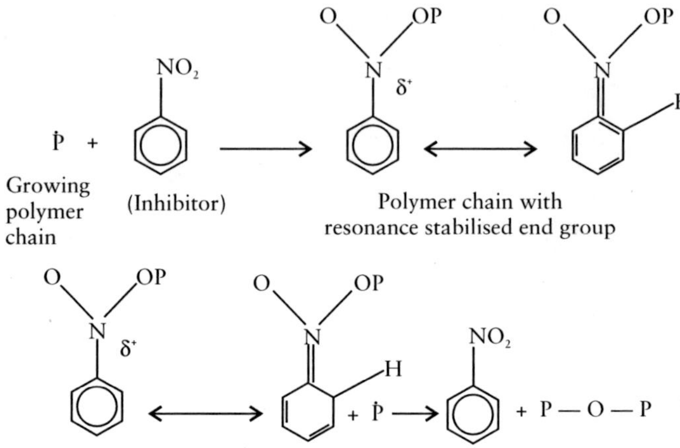

The nitro compound end of the chain is resonance stabilized, hence becomes less active. However, it can recombine with the radical of another growing chain to terminate the growth of the chain. In this way, one molecule of the inhibitor stops two growing chains.

The inhibitors are also used to stop the polymerization at a particular point. It is also added to the reaction mixture to avoid cross-linking in the polymer. The degree of polymerization is also checked by the use of the inhibitor.

Retarders: On the contrary, the substances which only reduce the rate of polymerization are known as retarders. Retarders decrease the concentration of free radicals as well as shorten the average life of the free radicals. Retarders like triphenyl methyl, diphenyl picrylhydrazyl are free radicals but are not sufficiently reactive to initiate the polymerization.

Basically, there is no difference between inhibitors and retarders, except the extent of effectiveness to control and stop the polymerization reaction. Inhibitors completely quench the reaction while retarders may increase the induction period or temporarily quench the polymerization reaction. A compound can act as an inhibitor in one polymerization or act as a retarder in another polymerization and may not affect at all a third polymerization.

For example, nitro compounds are inhibitors for vinyl acetate polymerization, retarder for styrene polymerization, but do not affect methylmethacrylate polymerization.

Atmospheric oxygen is a strong inhibitor for all polymerization reactions. The formation of biradical of oxygen is responsible for the inhibitory action of oxygen. Hence, the majority of polymerization reactions are carried out in an inert atmosphere of N_2.

The biradical oxygen reacts with the radical of growing chain a follows:

$$\dot{P} + O — O \rightarrow P — O — \dot{O}$$

$$P — O — \dot{O} + P \rightarrow P — O — O — P$$

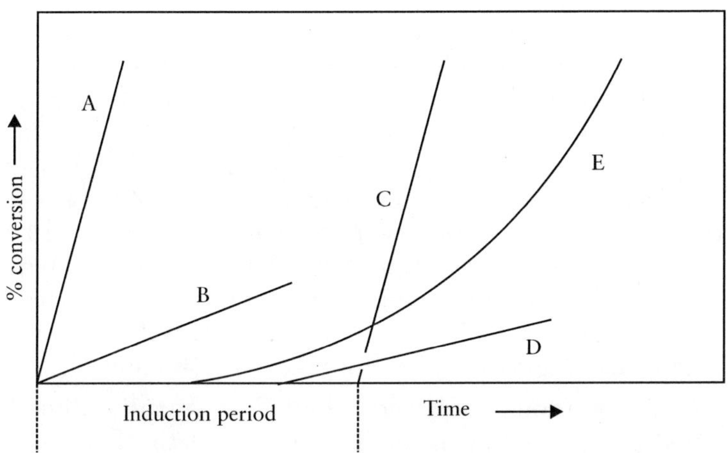

Fig. 2.1 Effect of inhibitors and retarders curve A: normal polymerization, curve, B: effect of retarder, curve, C: ideal inhibition, curve, D: inhibition followed by retardation, curve E: effect of non-ideal inhibitor. The time between A and C is induction period caused by an ideal inhibitor.

Nowadays about 40-45% of the plastic material and synthetic rubber are synthesized worldwide by free radical polymerization processes. Some of the important end-polymers that are synthesized by free radical mechanism are polyethylene, polyacrylonitrile, polystyrene, polyvinylchloride, polyisoprene, etc.

2.2.2 Ionic Polymerization

Ionic polymerization includes ionic initiation that can be either cationic or anionic. This depends on the nature of the carrier ion which is on the growing centre of constituting polymer chain. The cationic polymerization occurs by an electrophilic attack which undergoes an electrophilic addition to the monomer and renders the monomer a positively charged species. The reverse is true for anionic polymerization. It occurs by a nucleophilic attack and the monomer proceeds to polymerize by carrying a negative charge centre.

$$R^{\oplus} + CH_2 = CHX \longrightarrow R - CH_2CHX^{\oplus}$$

$$R^{\ominus} + CH_2 = CHX \longrightarrow R - CH_2CHX^{\ominus}$$

First, the formation of negative and positive charge centres on the monomer molecule is initiated. Then propagation takes place by the addition of more molecules of the monomer on this charged terminal successively. In contrast to termination reaction of free radical mechanism in cationic reactions terminations takes place by either unimolecular reactions or the active centre is transferred to other monomer molecule or solvent molecule or even to the molecules of impurity. Sometimes quenching of reaction is done by the addition of some reactive species to have a polymer of required molecular weight. While in anionic reactions the termination does not take place spontaneously.

2.2.2.1 Cationic polymerization: The broad outline of cationic polymerization is similar to that of the free radical polymerization, except that the initiator produces cations by donating a proton instead of producing free radicals. Generally, the catalyst requires a co-catalyst. For example, protonic acid catalysts such as BF_3, $TiCl_4$, $SnCl_4$, perchloric acid, tri fluoro-acetic acid, sulfuric acid, etc. react in the presence of co-catalyst like water, alcohol, acetic acid or acetyl perchlorate. These catalysts together with co-catalysts generate cations in the reaction mixture. Generally, monomers like isobutylene, styrene, vinyl allyl ethers, etc. are polymerized by cationic polymerization. This polymerization can be carried out at low temperature and proceeds at a fast rate so that the polymer of molecular weight as high as 10^6 is obtained quickly. The polymerization of isobutylene with catalyst boron trifluoride along with water as co-catalyst is given as follows:

Mechanism

1. Initiation: The catalyst and co-catalyst form a complex

$$BF_3 + H_2O \rightarrow H^+(BF_3OH)^-$$

Catalyst co-catalyst cation (complex)

This complex transfers a proton to the monomer, isobutylene, so that corresponding carbonium ion results in:

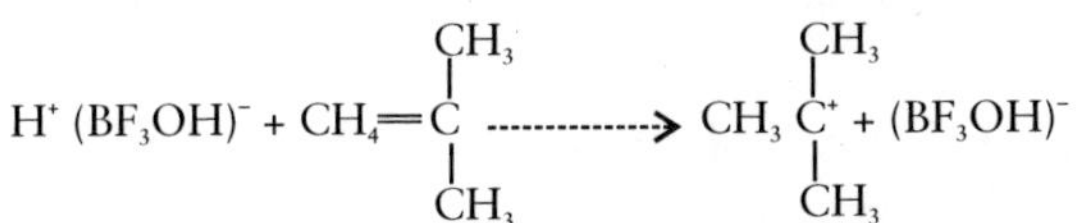

Cation monomer carbonium ion

2. Propagation: This carbonium ion reacts with another molecule of the monomer and chain growth occurs.

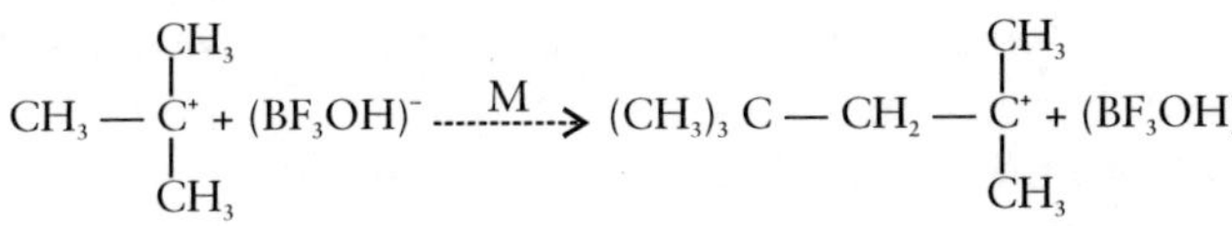

The general chain growth reaction can be shown as follows:

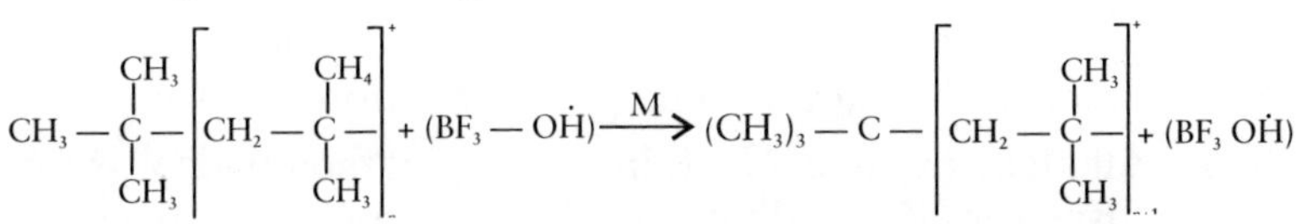

During the chain growth, the counterion (also called a gegen ion; gegen is a German word meaning opposite) always remains in the vicinity of the growth centre. This reaction is carried out in solvents of low dielectric constant, which have less amount of energy. The separation of ions needs a large amount of energy. As a result, the counter ions remain in the vicinity of the cationic centre.

3. Termination: The next step is termination or in some cases the chain transfer reaction, both of which are hindered by crowding of cationic centre.

Termination can take place in two ways.

(a) A polymer molecule with one unsaturated end unit is obtained by rearrangement of the ion pair.

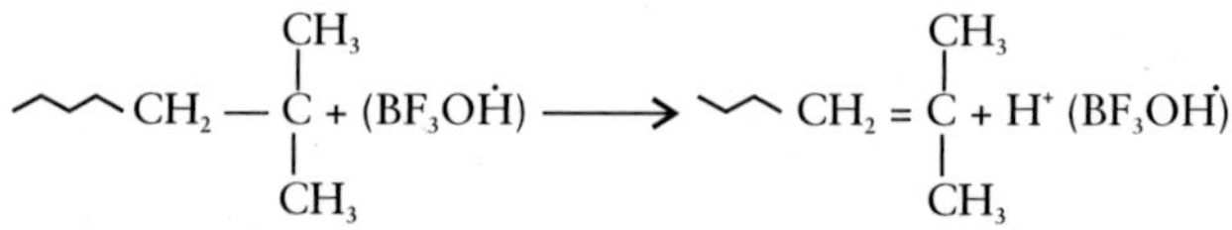

(b) A polymer molecule is obtained by transferring its charge to another monomer, which will make the catalyst separate out.

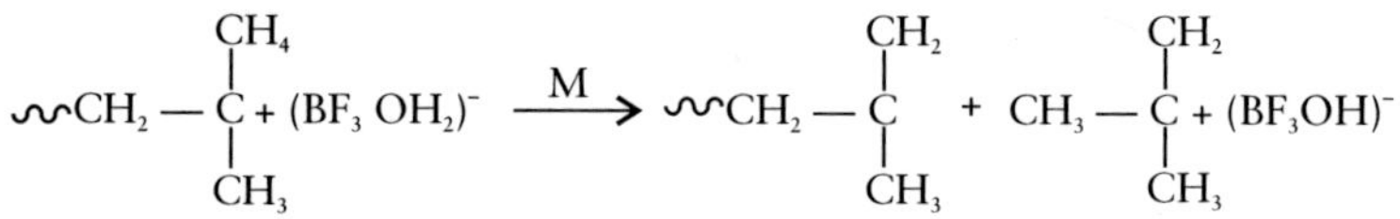

2.2.2.2 Anionic Polymerization: Anionic polymerization also takes place in three steps just as it happens in other addition polymerizations, viz. (1) initiation, (2) propagation and (3) termination. The monomers that have electronegative substituents like —CN, —COOR, —C_6H_5, etc. are polymerized by anionic polymerization since the electronegative substituent facilitate to form stable anions, i.e. carbanion, by pulling out an electron from the double bond. This carbanion is susceptible to the attack by an electron donor. A co-catalyst is also needed in anionic polymerization, e.g. organic solvents like heptanes. In anionic polymerization, no chain transfer or branching occurs. The initiating entities for anionic polymerization are:

1. Alkali metals (e.g. Na, K in the form of their hydrocarbon suspensions) or alkali metal complexes with aromatic compounds, liquid ammonia, ether, etc.
2. Lewis bases (e.g., ammonia, primary amines, aniline).
3. Organometallic compounds (e.g. butyl lithium, Grignard reagent).
4. High-energy radiation.

The initiation reaction is very fast in anionic polymerization. The initiation can take place in two ways:

(1) by direct attack of the base to monomer to form corresponding carbanion (anion) and

(2) by transferring an electron from the donor molecule to the monomer to form a carbanion (anion).

Lewis bases and organometallic compounds initiate the polymerization by direct attack and alkali metals/alkali metal complexes initiate by transferring an electron.

As an example, the polymerization of styrene with sodium metal in the presence of naphthalene is given as follows:

(1) Initiation

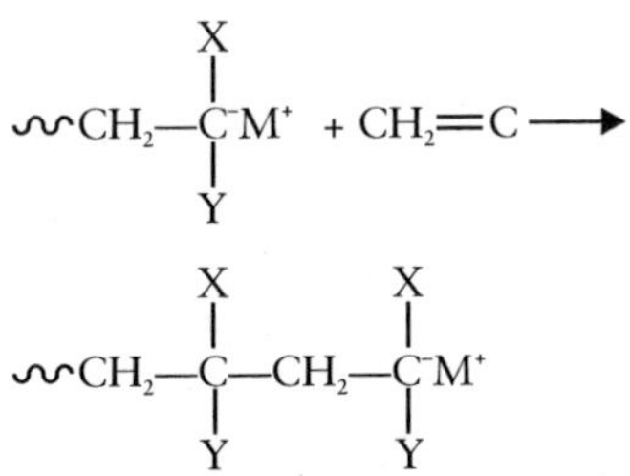

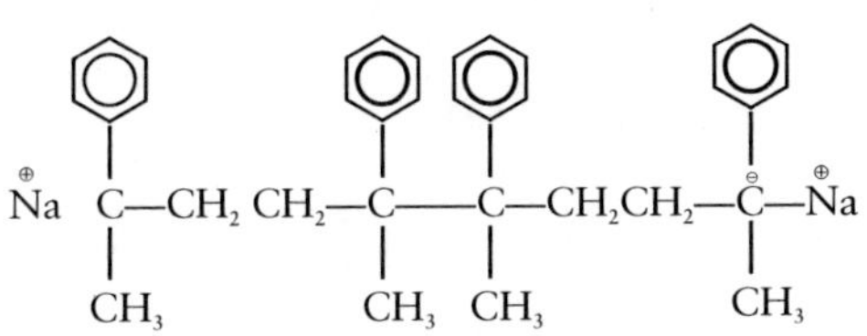

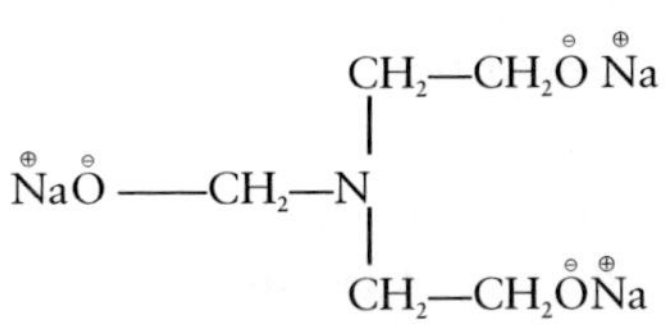

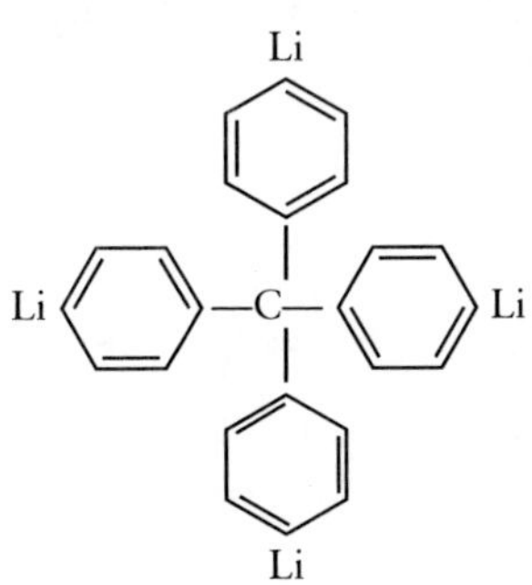

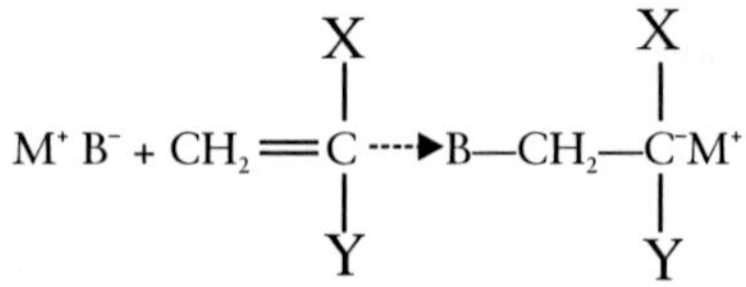

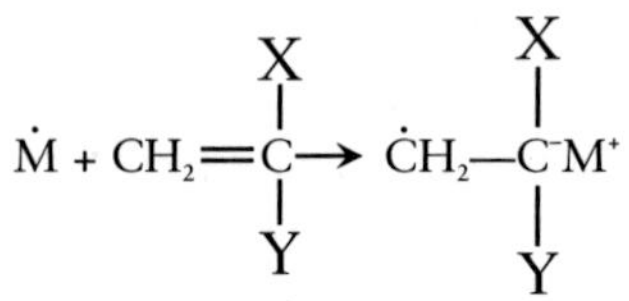

Anion Radical

Initiation by $NaNH_2$ in NH_3 takes place as follows:

$$NaNH_2 \xrightarrow{NH_3} Na^+NH_2^-$$

Initiation can be a two-way, three-way or four-way growth depending on the choice of the initiator.

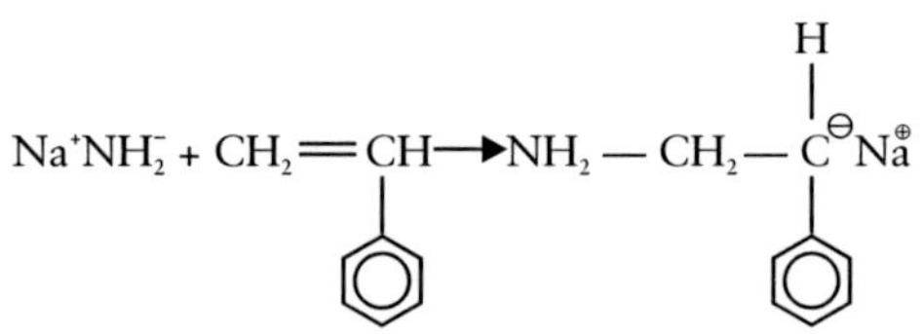

Different initiators like metal amide, alkoxide, alkyl, aryl and hydroxide have been represented by a general formula M^+B^-. The choice of initiator depends on the basicity of initiator and the acidity of the monomer, e.g. the polymerization of styrene cannot take place with OH^- whereas *tris* (sodium ethoxy) amine $N(CH_2CH_2O^-\ Na^+)_3$ is able to initiate growth of three chains simultaneously.

(2) Propagation

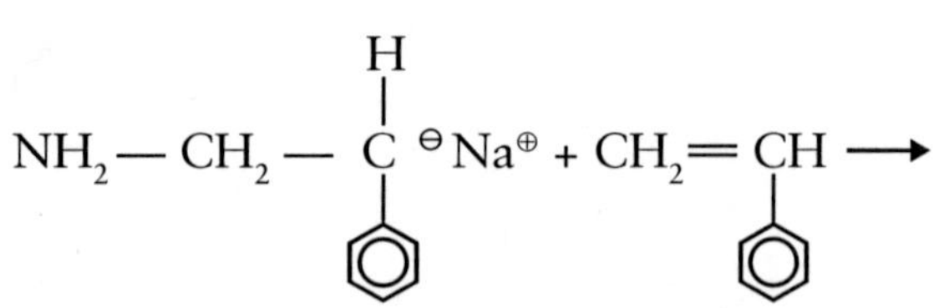

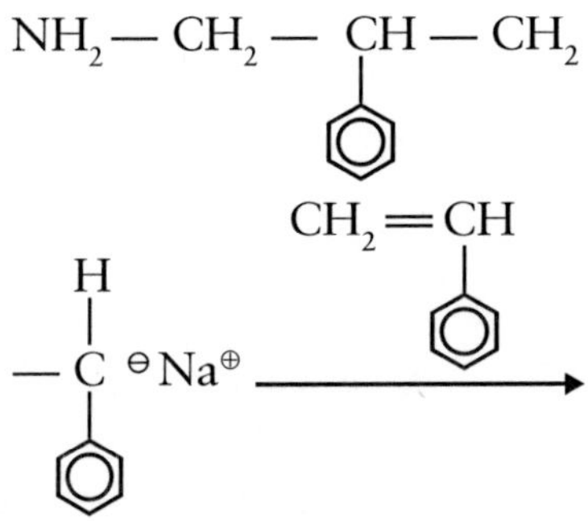

Propagation step is considered as the rate determining step because it is much slower than the initiation step. The propagation reaction depends considerably on the nature of the gegen ion. The monomer molecule adds to the growing chain by squeezing itself between the chain and the gegen ion as a result of which resonance, polar, and steric effects play a significant role in determining kp.

(3) Termination

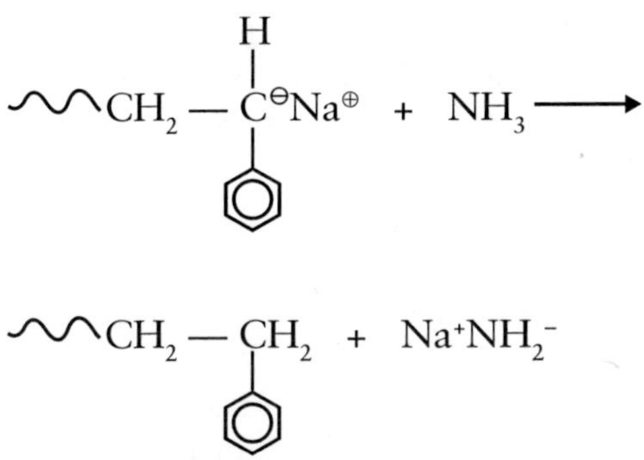

Termination of the polymer chain does not take place automatically as it occurs in other types of polymerizations. It occurs when molecules that are capable of reacting with the active chain ends like oxygen, carbon dioxide, methanol, water, etc. are either deliberately or accidentally introduced into the system.

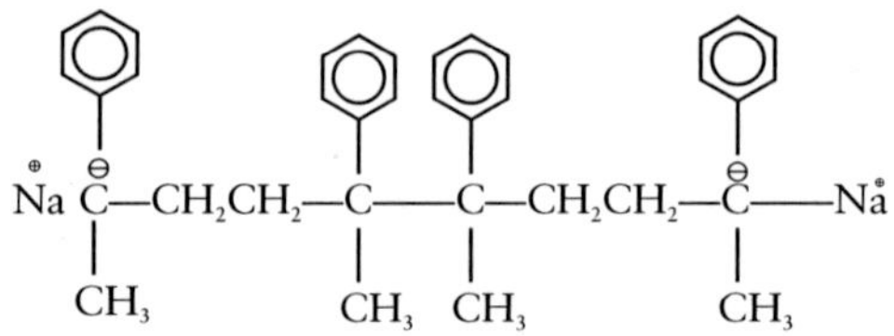

The nature of gegen ion in anionic polymerization affects significantly the growth of the polymer chains.

In some anionic polymerizations, if the reactants are pure and there are no traces of oxygen and water in the system, the termination does not take place. The active polymer molecules remain in the reaction mixture, even after the consumption of monomer molecules, so that the polymerization can be resumed by the addition of fresh monomer. Such polymeric molecules are called "living polymers". Since throughout the reaction, the chain-ends grow at the same rate, it is possible to determine the molecular weight of living polymers by the following ratio (monomer concentration to that of the initiator).

$$\text{Degree of polymerization (DP)} = \frac{[\text{Monomer}]}{[\text{Initiator}]} \qquad ...(2.1)$$

The main characteristic feature of the polymers produced by living polymerization is that they have very narrow molecular weight distribution.

2.2.2.3 Coordination Polymerization: One of the major steps in the field of polymer chemistry was taken in 1953 with the discovery of new catalysts that lead to the formation of stereo-regular polymers. These catalysts are known as Ziegler-Natta catalysts and the polymerization is known as Ziegler-Natta Coordination Polymerization or Coordination Polymerization. Coordination polymerization is also known as Stereoregular Polymerization. It is carried out in the presence of special catalyst-co-catalyst systems, popularly called Ziegler-Natta catalysts. Ziegler-Natta catalysts involve the reaction of halides of the transition metals of group IV-VIII in their lower oxidation state (e.g. Ti, V, Cr, Zn), with organo-metallic compounds (e.g. alkyls, aryls or hydrides) of group I-III metals (e.g. Al, Mg, Li).

The polymerization of olefines and dienes catalyzed by organo-metallic compounds falls under the category of Zeigler-Natta polymerization. This polymerization method does not need high temperature and pressure and takes place under mild conditions. The most striking feature of this polymerization is the high degree of control it has on the structure of the polymer formed. There are three phases in the reaction mixture of co-ordination polymerization:

(i) The monomer

(ii) The solvent

(iii) The catalyst

The polymerization reaction takes place on the surface of catalyst particles. The incoming monomer is added at the junction of the catalyst and already present monomer. The first step is the formation of the monomer-catalyst complex. For example,

(1) Alkyl aluminium and $TiCl_3$

(2) The reaction of $TiCl_4$ with tri-Alkyl aluminium produces titanium tri-cloride and various alkyl metal chlorides, which is represented as follows:

$$TiCl_4\ AlR_3 \rightarrow RTiCl_3 + R_2AlCl$$

$$TiCl_4 + R_2AlCl \rightarrow RTiCl_3 + RAlCl_2$$

$$TiCl_4 + RAlCl_2 \rightarrow RTiCl_3 + AlCl_3$$

$$RTiCl_3 \rightarrow R + TiCl_3$$

The composition of these catalysts is not well defined, but the most accepted structures are either alkylated metal halide (mono-metallic structure I) or bi-metallic complex, among which the latter has bridge between the two metals (structure II)

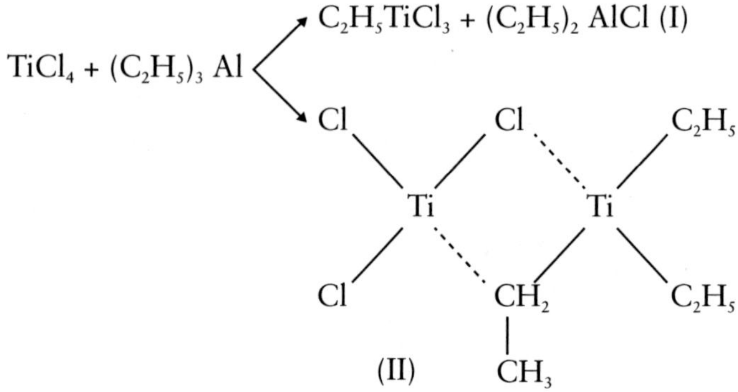

Co-ordination polymerization may follow the anionic or cationic path, but the examples of cationic co-ordination polymerization are fewer. Only the monomers containing hetero-atoms with a lone pair of electrons like oxygen, nitrogen, etc., undergo cationic co-ordination polymerization. In general, the growth of the polymer chain in co-ordination polymerization is controlled by the counter ion of the catalyst and plays the key role in the polymerization. First, it coordinates and simultaneously orients the incoming monomer. The catalyst then inserts the polarized double bond of the monomer in the already polarized bond between the counter ion and the end of the growing chain. The overlapping of the electrons of monomer takes place either with the empty sp-orbital of group I-III metals or with the empty d-orbital of transition metal, as is shown in the following mechanism.

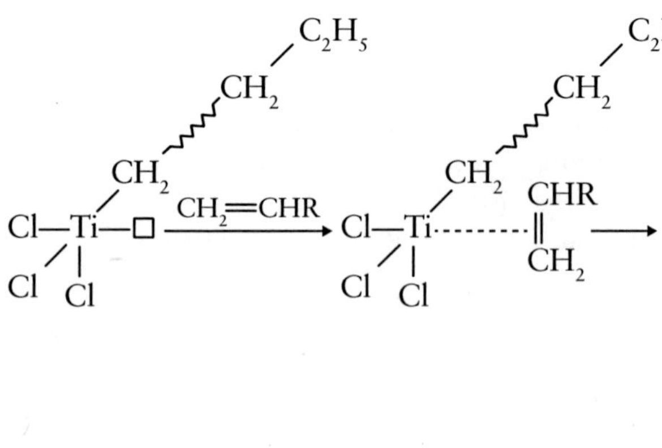

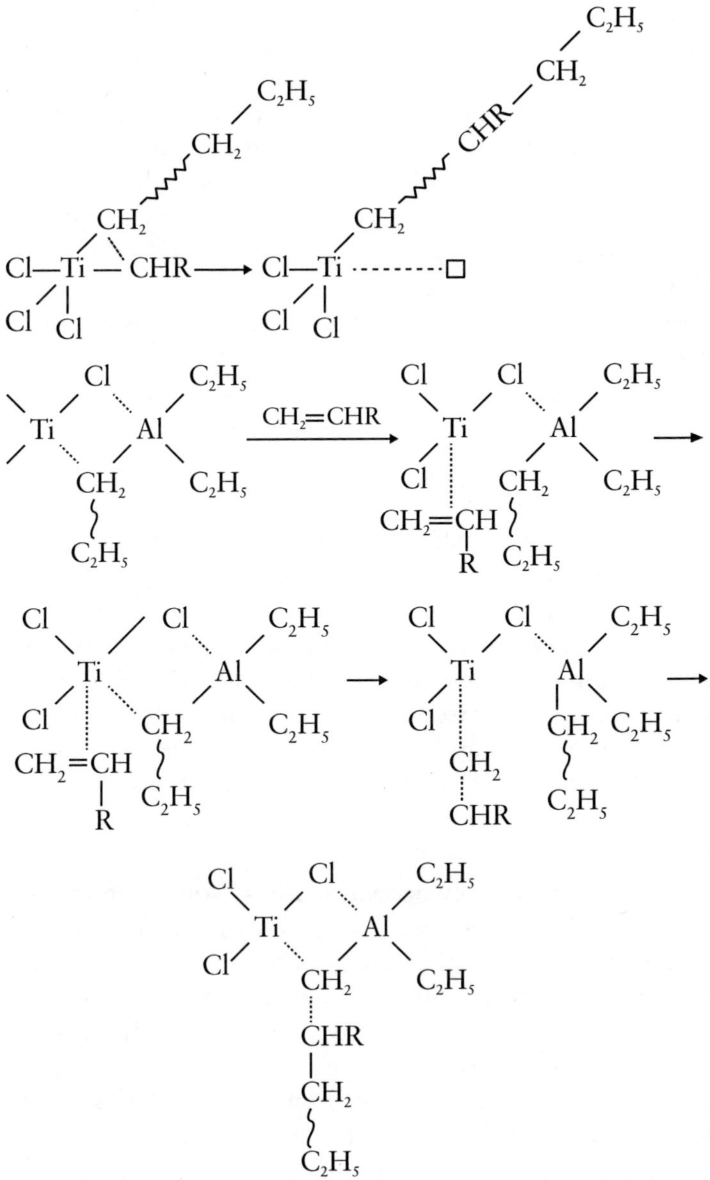

In the termination step, the chain can either be transferred to the transition metal or to the group I-III metal alkyl or to an active hydrogen compound like molecular hydrogen.

Ziegler-Natta polymerization can be carried out in heterogeneous as well as in homogenous phase. When it is carried in homogenous phase with a non-polar monomer, mostly atactic polymer and sometimes syndiotactic polymer is obtained. But when carried out in heterogeneous phase, an isotactic polymer is obtained. Ziegler-Natta polymerization gained much importance because it can synthesize atactic polymer with the help of non-polar monomers.

In heterogeneous reaction, the catalyst is insoluble in the solvent. But, if the catalyst is homogenous, i.e. it is soluble in the solvent, then it loses the stereo-chemical control.

The kinetics of Ziegler-Natta polymerization is complicated because of the heterogeneous nature of the reaction.

Ziegler-Natta polymerization has characteristics of a living polymerization, but with reference to its active site only. The life of active site is much less as compared to the life of the propagating chain.

2.2.2.4 Ring-Opening Metathesis Polymerization: When in a cyclic molecule the ring structure is strained, then it can act as a monomer for polymerization as its ring can be opened up and if conditions are favourable, it can be polymerized. The more strained four- and eight-membered rings are polymerized easily. While six-membered rings are less strained and hence, on polymerization, yield less product. However, five-membered rings are difficult to polymerize.

Ring-opening polymerization has some similarities as well as dissimilarities with step polymerization and chain polymerization. In the ring-opening polymerization, no small molecule is liberated during the reaction unlike in step polymerization. In the same way, it retains unsaturation unlike in chain polymerization. This polymerization has features of both, chain and step polymerization as far as their kinetics and mechanism are concerned. It resembles chain polymerization, since the reaction proceeds by the addition of monomer units. Its resemblance with step polymerization is shown since the polymer growth occurs by stepwise addition of monomer units so that the molecular weight of the polymer increases gradually. This polymerization generally results in linear polymers. A number of compounds are polymerized by this method, e.g. epoxides (ethylene oxide, propylene oxide) cyclic ethers (trioxane, tetrahydrofuran), cyclic esters (lactones), cyclic amides (lactames), cyclo-olefines and siloxznes. But all cyclic compounds do not undergo ring-opening polymerization.

The mechanism of ring-opening metathesis polymerization can be understood by taking the following examples:

Cyclic Ethers: Cyclic ethers which are also known as epoxides, e.g. ethylene oxide can be polymerized in several ways, e.g. by Lewis acids, by anionic polymerization. The choice of a method depends on several factors, viz. ease of method, to get the end-product of desired characteristics, etc. However, every method has main three stages: (1) initiation, (2) propagation and (3) termination. As an example, we see the cyclic polymerization of ethylene oxide (a symmetrical epoxide) by a base M^+A^-.

(1) Initiation

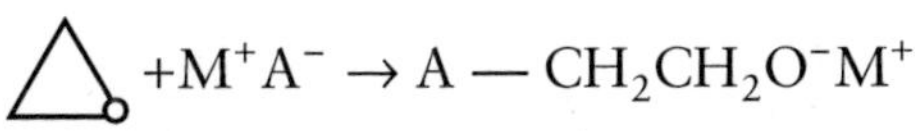

Ethylene oxide initiator

(2) Propagation

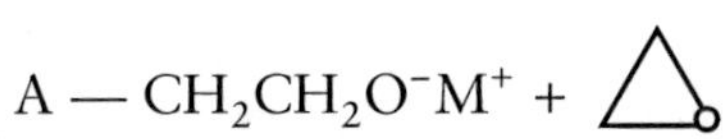

$$\rightarrow A - CH_2CH_2OCH_2CH_2O^-M^+$$

$$A \left(CH_2CH_2\,O \right)_n CH_2CH_2O^-M^+$$

(3) Termination

In many of these polymerizations, there is no termination reaction and thus a living polymer is obtained. For terminating the reaction, a terminating agent is to be added externally.

In the case of polymerization of an unsymmetrical epoxide like propylene oxide, the ring-opening can occur in more than one ways so that different polymers can be obtained. As is shown in the following equation:

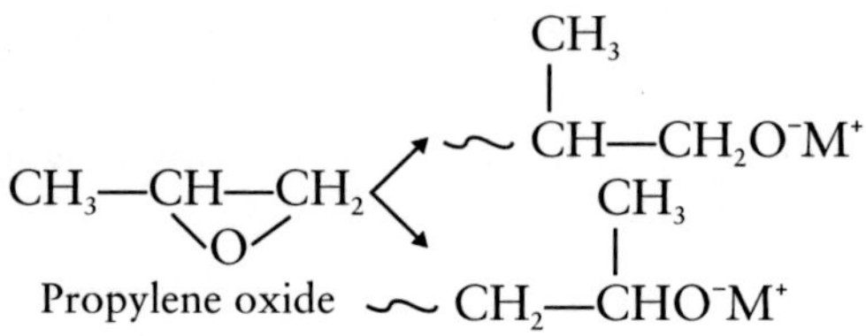

Cyclic Amides (Lactams)

The ring-opening of lactams is initiated by water and anionic reagents to get the polymer as follows:

$$n\ \overline{CO—NH—(CH_2)_5} \xrightarrow[\Delta]{H_2O} [NH—(CH_2)_5—CO]_n$$

In this type of polymerization, three main equilibria are involved

(1) Hydrolysis

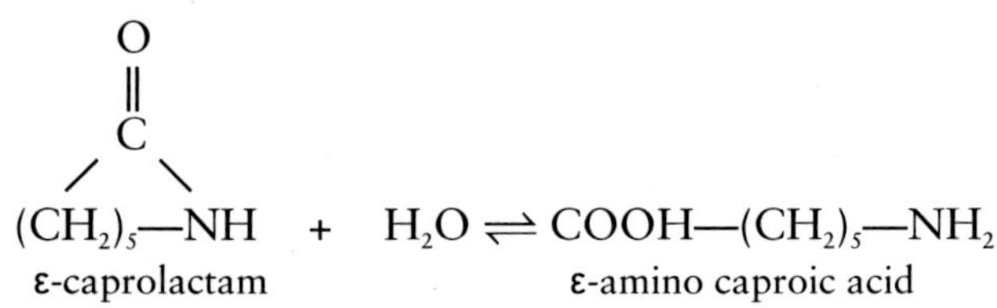

(2) Step polymerization

$$COOH—(CH_2)_5—NH_2 +$$
ε-amino caproic acid

$$COOH—(CH_2)_5—NH_2 \rightleftharpoons H_2O +$$
ε-amino caproic acid

$$COOH + CH_2 +_5 NHCO(CH_2)_5 NH_2$$

(3) Ring-opening polymerization

(i) Initiation

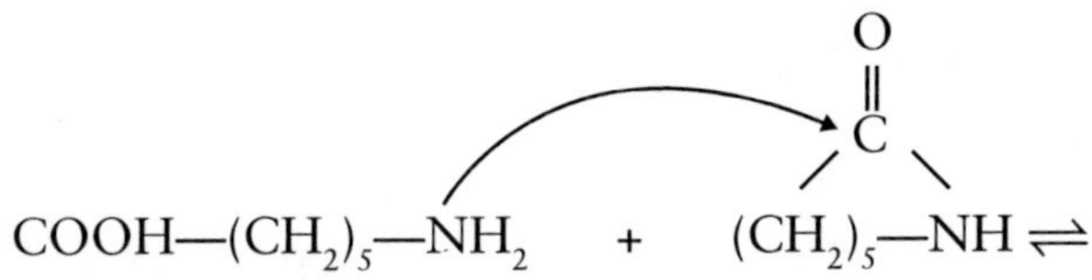

$COOH—(CH_2)_5—NH\ CO\ (CH_2)_5\ NH_2$

(ii) Propagation

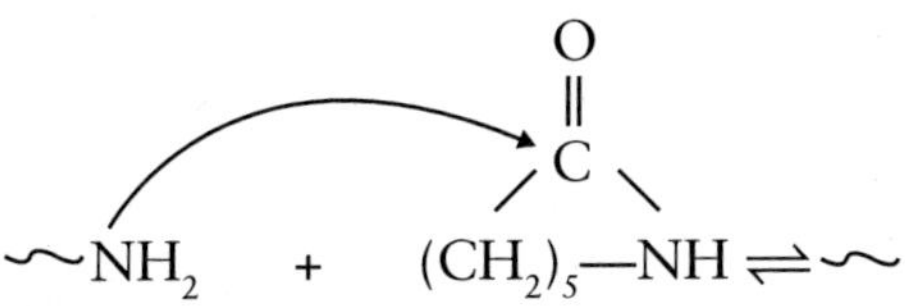

$NHCO(CH_2)_5NH_2$

The propagating reaction is predominantly a ring-opening rather than the step polymerization of the caprolactam.

Strong bases (B^-M^+) like alkali metals or sodium methylate form the leactam anion can to set off the polymerization of cyclic amides. A tentative scheme of this polymerization is given below:

(1) Initiation

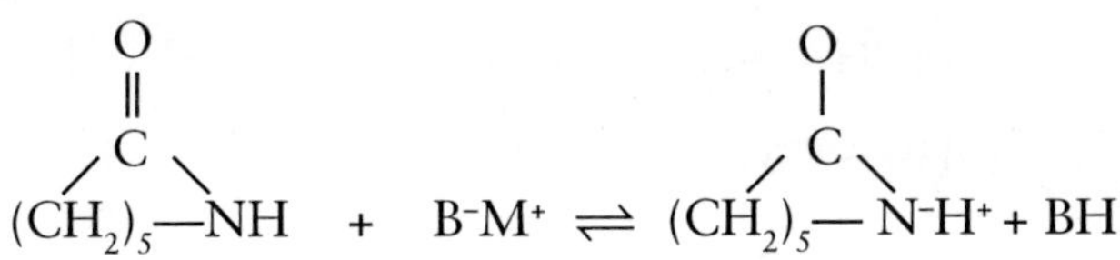

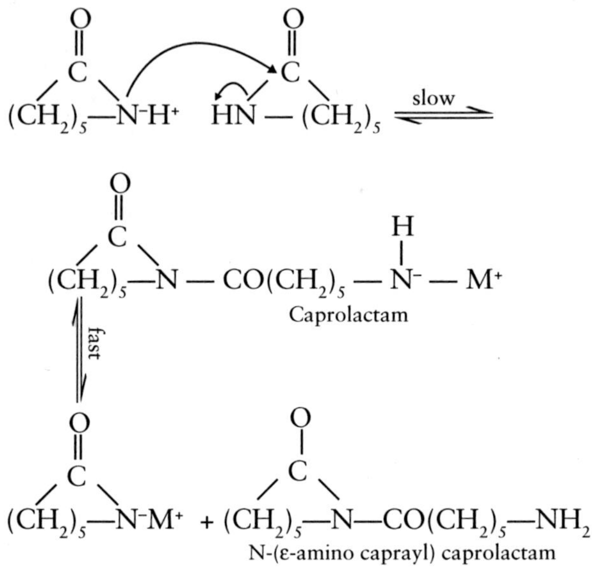

(2) Propagation

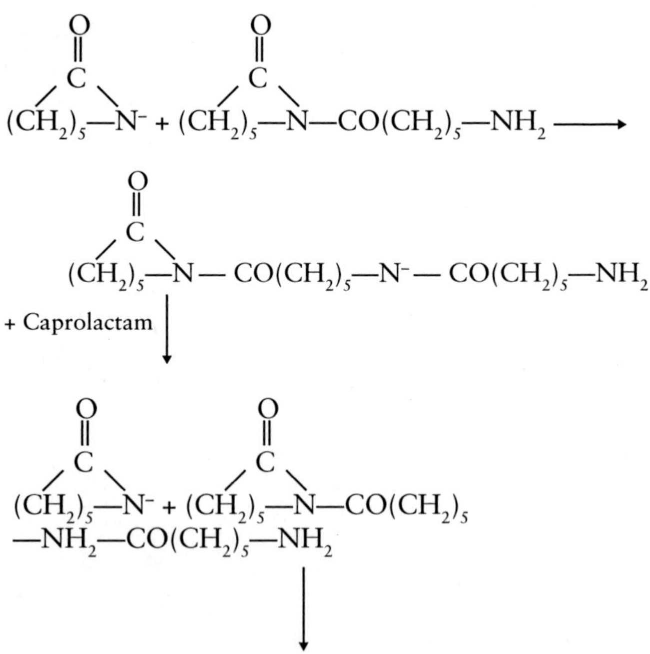

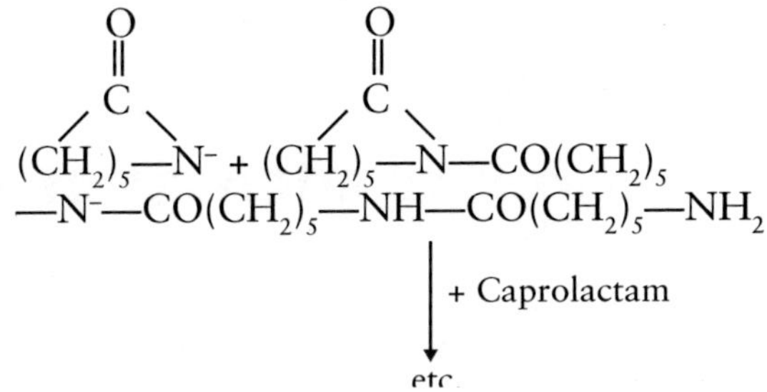

(3) Termination

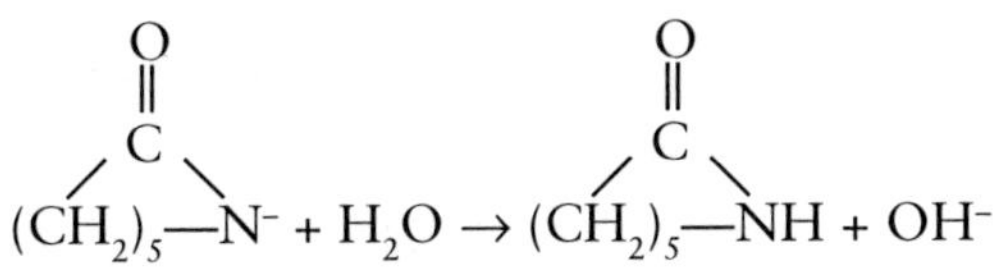

Always caproycaprolactam ring at the end of the chain is obtained.

2.3 Phase Systems in Polymerization (Techniques of Polymerization)

The chemical process of a polymerization can be divided into three stages, viz. chemical reaction, separation (or purification) and identification. The most complex one among these is the separation in terms of time and resources. The polymer gels have gained significant importance in heterogeneous catalysis as they fail to dissolve in any given medium and the separation step reduces to the simple removal of various reacting fluids. Recently, a new phase called the fluorous phase has been discovered which is immiscible in both organic and aqueous phases. But as their synthesis cost is very high, hence, they are used only for research purposes. Conceptually, this approach is analogous to solid-phase separation, however, there is a difference that here it is in a liquid state.

3 Step Growth Polymerization (Polycondensation)

As the name implies, these polymers are made by a long series of sequential steps that take a long time to form a high molecular weight polymer by the formation of a series of low molecular weight intermediates, commonly known as dimers, trimers, tetramers, pentamers, etc. in general, they are known as oligomers (unlike chain growth). For quite a long time the reaction mixture has oligomers of various molecular weights along with monomer molecules. The reaction takes place in such a manner that it avoids the formation of a ring or cyclic structures and proceeds indefinitely. If polymerization takes place using bifunctional monomers, then a linear polymer is formed. If the polymerization is carried out with trifunctional monomers then a non-linear polymer occurs through the formation of branched polymer. Functional groups or atoms that react by step polymerization reaction are: reactive hydrogen (—H), hydroxyl (—OH), halogen (—Cl, —Br), carboxyl (—COOH), amino (—NH_2), isocyanate (—NCO), aldehyde (—CHO) and double bond (C = C), etc. The sequences of reactions are normally associated with the formation of a condensed inter-unit chemical link and liberation of a small by-product such as H_2O, NH_3, alcohol, CO_2 and HCl, etc. at each step. There are many types of condensation polymers some important are polyesters, polyamides, polyethers, polyurethanes, etc.

Polyesters: They are synthesized by condensation of a dicarboxylic acid with a diol. A linear polyester can also be prepared by self-condensation of a hydroxycarboxylic acid.

$$\text{nHO}-\text{R}-\text{COOH}-\text{nHO}-\text{R}-\text{COOH} \longrightarrow$$

(Hydroxy acid)

$$\text{H}\!+\!\text{O}-\text{R}-\dot{\text{C}}\text{OO}-\text{R}-\text{CO}\!+_n \text{OH}$$

[Poly (ester)]

$$\text{nHOOC}-\text{R}-\text{COOH}-\text{nHO}-\dot{\text{R}}-\text{OH} \longrightarrow$$

Dicarboxylic acid

$$\text{HO}\!+\!\text{CO}-\text{R}-\dot{\text{C}}\text{OO}-\dot{\text{R}}-\text{O}\!+_n \text{H}$$

[Poly (ester)]

Polyamides: They are obtained by the condensation of a diamine with a dicarboxylic acid.

$$\text{nH}_2\text{N}-\text{R}-\text{NH}_3 + \text{nHOOC}-\dot{\text{R}}-\text{COOH} \longrightarrow$$

(Diamine) Dicarboxylic acid

$$\text{H}\!+\!\text{HN}-\text{R}-\text{N}\dot{\text{H}}\text{CO}-\dot{\text{R}}-\text{CO}\!+_n \text{OH}$$

Polyethers: They are obtained when an epoxide group, containing compound reacts with a diol.

$$\underset{\text{Epoxy derivative}}{\overset{\;\;\text{O}}{\text{CH}_2-\text{CH}}-(\text{CH}_2)_n\,\text{Cl}} + \underset{\text{Glycol}}{\text{OH}_2-\text{R}-\text{OH}} \longrightarrow$$

$$-\text{O}-\text{CH}_2-\underset{\text{OH}_2}{\underset{|}{\text{CH}}}-(\text{CH}_2)_n-\text{O}-\text{R}-\text{O}$$

(Polyether)

Polyurethanes: They are synthesized by the condensation of a diol with a diisocyanate.

$$\underset{\text{Glycol}}{\text{HO}-\text{R}-\text{OH}} + \underset{\text{Disocyanate}}{\text{OCN}-\dot{\text{R}}-\text{NOO}} \longrightarrow$$

$$+\!\text{O}-\text{R}-\text{O}-\underset{\text{OH}_2}{\underset{\|}{\dot{\text{C}}}}\text{NH}-\text{R}-\text{NH}-\underset{\text{O}}{\underset{\|}{\text{C}}}\,)_n$$

[Poly (urethane)]

As a general reaction, we take the example of condensation of a dicarboxylic acid and a diol.

Step 1:

$$\text{HO}-\text{R}-\text{OH}+\text{HOOC}-\text{R}'-\text{COOH} \rightarrow \text{HO}-\text{R}-\text{OOC}-\text{R}'-\text{COOH}+\text{H}_2\text{O}(\text{I})$$

Step 2:

$$\text{HO}-\text{R}-\text{OOC}-\text{R}'-\text{COOH}+\text{HOOC}-\text{R}'-\text{COOH} \rightarrow$$

$$\text{HOOC}-\text{R}'-\text{CO}-\text{O}-\text{R}-\text{OOC}-\text{R}'-\text{COOH}+\text{H}_2\text{O}\ (\text{II})$$

Or

$$HO—R—OOC—R'—COOH + HO—R—OH \rightarrow HO—R—OOC—R'—COO—R—OH + H_2O \text{ (III)}$$

Or

$$HO—R—OOC—R'—COOH + HO—R—OOC—R'—COOH \rightarrow HO—R—OOC—R'—COO—R—OOC—R'—COOH \text{ (IV)}$$

A diol molecule reacts with a dicarboxylic acid molecule to form a dimer. This dimer has —OH group at one end and —COOH group at the other. Now there are three possibilities for further reaction: (a) the dimer can react with another molecule of the dicarboxylic acid to form a tetramer having —COOH groups on both ends, (b) or the dimer can react with the diol to form a tetramer having —OH groups on both ends, and (c) lastly, it can react with another molecule of dimer to form a tetramer with —OH group at one end and —COOH group at the other end. In this way, the reaction goes on to form a long chain. The condensation products of all the successive steps retain the bifunctional character. The reaction of a long-chain intermediate with an unreacted monomer molecule is always possible.

Production of a polymer molecule of chain length 'n' involves completion of (2n – 1) successive steps of condensation and can be represented as follows:

$$nHO—R—OH + nHOOC—R'COOH \rightarrow HO[—R—OOC—R'—COO]_n H + (2n-1)H_2O$$

Step polymerization can be looked at as a simple condensation reaction, which is carried out repeatedly, e.g. polyesterification reaction to synthesize terylene is same as a simple esterification reaction between alcohol and carboxylic acid. The equilibrium of this reaction can be shifted to the forward direction by the removal of the by-product, viz. water, either by simple distillation or is removed as an azeotrope. After some time the molecular weight of the product increases, which in turn increases the viscosity of the reaction mixture and finally the polymer precipitates. Hence, powerful stirrers and strong vessels are required when the synthesis is carried out on a large scale.

The polycondensation reaction can be stopped at any time and low molecular weight polymers (either with —OH or —COOH group as terminal group) can be isolated. This is in strong contrast to chain polymerization in which the reaction cannot be stopped to get intermediates.

We have already understood in section 1.3 that polyfunctionality or, at least, bifunctionality is the necessary condition for step growth polymerization. On the basis of functionality of monomer, there can be two classes of step-growth polymerization and hence step-growth polymers.

1. Polymers can be synthesized by two monomers, each having the same functional groups its ends. For example, the adipic acid has two terminal —COOH groups. Another monomer is hexamethylenediamine with two amino groups at both the terminals. These two bifunctional monomers react together to form polyamide, which is a well-known step-growth polymer. Another example is the formation of polyesters formed by reaction between diols and dicarboxylic acids.

(1) On the basis of reactant monomers having same bifunctional groups, i.e. bifunctional monomer A-A and bifunctional monomer B-B. It can further be divided into two types

1. Polyesters: These are formed by reactions between diols and dicarboxylic acids.

$$\mathrm{nHO—R—OH + nHOOC—R'-COOH \rightarrow (—O—R—O—OC—R'—CO—)_n + 2nH_2O}$$

2. Polymides: These are formed by reactions between diamines and dicarboxylic acid.

$$\mathrm{nH_2N—R—NH_2 + nHOOC—R'—COOH \rightarrow (—NH-R—NH—OC—R'—CO—)_n + 2nH_2O}$$

In both the classes (1) and (2), the hydrocarbon sequences R and R′ can be same or different, it can contain any number of carbon atoms and can be linear or cyclic.

(2) The second type of polymers is formed by monomers with two different functional groups on the same molecule, for example, the monomer p-aminobenzoic acid in which both, amino and carboxylic acid groups are present on the same molecule. Chain growth proceeds from reactions between two different functional groups belonging to different monomer molecules. However, there are exceptions to this generalization like in the formation of polyanhydride from a single monomer having one type of groups, viz. both are carboxylic groups and throwing out one water molecule during condensation reaction. This class can further be divided into two:

Polyaminoacid

$$\mathrm{nH_2N—CR—COOH \rightarrow (—NH—CR—CO—)_n + nH_2O}$$

General

$$nAB \rightarrow \mathrm{(AB)}_n + ...$$

The classes (1) and (2) involve hydrocarbon sequence between the reactive groups A and B. However to understand the step-growth reactions class (2) 2. is considered as a prototype.

3.1 Mechanism of Step Growth Polymerization

In step-growth polymerization, the functionality of the monomer is of prime importance. Simplest case is when monomer is a bifunctional compound. These reactions are analogous to simple reactions involving monofunctional units, e.g. esterification reaction of a diol and diacid.

$$\underset{\text{Diol}}{\mathrm{HO-R—OH}} + \underset{\text{diacid}}{\mathrm{HOOC—R'—COOH}} \rightarrow \underset{\text{ester}}{\mathrm{HO-R—OCO—R'—COOH}} + \mathrm{H_2O}$$

If water is removed from the reaction mixture the equilibrium is disturbed and as a result, the reaction proceeds in the forward direction. The resulting dimer is again bifunctional. This reacts with another molecule of acid from one end and from the other end with a base, forming trimers, tetramers and so on. All these are independent reactions.

Consider the synthesis of polyester from a diol and a diacid ateach step. The first step is the reaction of the diol and diacid monomers to form dimer:

$$HO—R—OH + HOOC—R'—COOH \rightarrow$$
$$HO—R—OCO—R'—COOH + H_2O$$

The dimer then reacts to form trimer with diol monomer

$$HO—R—OCOR'—COOH + HO—R—OH \rightarrow$$
$$HO—R—OCO-R'—COO—R—OH + H_2O$$

and also with diacid monomer:

$$HO—R—OCO—R'—COOH + HOOC—R'—COOH \rightarrow$$
$$HOOC—R'—COO—R—OCO—R'—COOH + H_2O$$

If two dimers react together tetramer is formed:

$$2HO—R—OCO—R'—CO_2H \rightarrow$$
$$HO—R—OCO—R'—COO—R—OCO—R'-CO_2H + H_2O$$

The tetramer and trimer react with each other and then with monomer and a dimer. Like this polymerization proceeds in a stepwise manner with a continuous increase in the molecular weight of the polymer with time (conversion). Step polymerization is characterized by the disappearance of monomer at an early stage in the reaction. Every molecule of monomer is consumed far before the production of an end-polymer of sufficiently high molecular weight (approximately > 5000-10,000). Thus, in the case of most step polymerizations, there remains less than 1% of the original monomer at an early stage of the reaction, where as the average polymer chain contains only 10 monomer units. However, the situation is quite different in the case of the chain polymerization.

The rate of step polymerization is, therefore, the sum of the rates of reaction between the molecules of various sizes. In other words, it is the sum of the rate of reactions such as

Monomer + monomer → dimer

Dimer + monomer → trimer

Dimer + dimer → tetramer

Trimer + monomer → tetramer

Trimer + dimer → pentamer

Trimer + trimer → hexamer

Tetramer + monomer → pentamer

Tetramer + dimer → hexamer

Tetramer + trimer → heptamer

Tetramer + tetramer → octamer

Pentamer + trimer → octamer

Pentamer + tetramer → nonamer

All the above reactions and more can be expressed as a general reaction:

$$n - mer + m - mer \rightarrow (n + m) - mer$$

Thus, the products as each step are again bifunctional compounds and the sequence of condensation reaction goes on. Like this, the polymer chain keeps growing step by step. The reaction mixture at any instance of time consists of various-sized diol, diacid, and hydroxy acid molecules. Any HO-containing molecule can react with any COOH-containing molecule. This is a general characteristic of step polymerization.

4 Copolymerization

4.1 Introduction

To have a polymer with a desired property, one can either mix two or more polymers to make a mixture. But the resulting mixture may not always be of good quality. The other method would be to synthesize a polymer chain containing more than one monomer/monomer-pair, which is known as copolymerization. By choosing two monomers or a monomer pair polymerized simultaneously, the resulting copolymer has more than one repeat unit. The resulting copolymer is observed to have better characteristics than a homopolymer, which is obtained from one monomer. If one considers the simplest case of two monomers reacting together, then a variety of structures can be obtained, which can be classified as follows:

1. **Statistical (Random) Copolymer:** This is the most commonly formed copolymer at the time when irregular propagation occurs, e.g. —ABAABBBAABBBABB—

2. **Alternating Copolymer:** When equimolar quantities of both the monomers are distributed in an alternate manner in the chain, they are known as alternating copolymer.

—ABABABA—

3. **Block Copolymer:** When chain propagation takes place so that a long sequence of one monomer is followed by a long sequence of the second monomer, then a block copolymer is obtained.

—AABBBAABBB—

4. Graft Copolymer: In graft copolymers, the chain of one monomer is attached to the main chain of another homopolymer. Hence, graft copolymers are also termed as branched block copolymers.

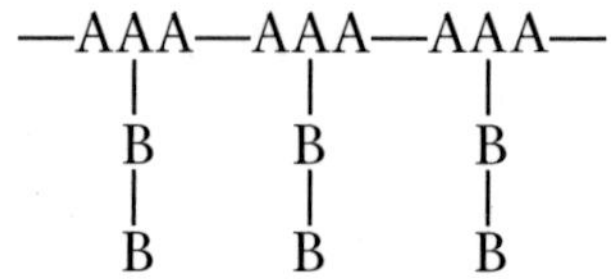

5. Stereoblock Copolymer: These copolymers have a special structure having tacticity. This is formed from one monomer.

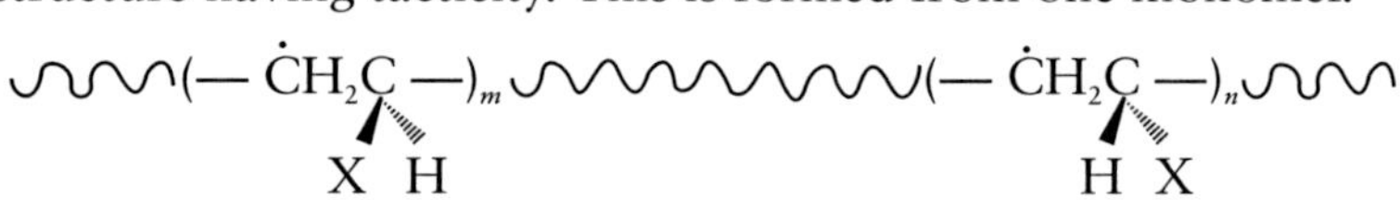

Fig. 4.1

It is worth mentioning here that the sequence distribution of different CRU in the copolymer chains is an important factor that controls the properties of copolymer. The sequence distribution, in turn, is affected by the monomer reactivity. The block copolymers and graft copolymers have properties of both the monomers by which they are formed. While alternating copolymers and random/statistical copolymers have properties in between the properties of both the monomers. The copolymerization mechanism is more complex than simple polymerization mechanism. There are many factors that affect copolymerization mechanism. These introduce complexity to copolymerization, e.g. the copolymer resulted from copolymerization of styrene and vinyl acetate has only 1-2% of vinyl acetate, whereas if the quantity of styrene used is small, it tends to inhibit the free-radical polymerization of vinyl acetate. Maleic anhydride and stilbene cannot be polymerized independently, but when these two are together they get copolymerized quite easily.

The general representation of random copolymerization is as follows:

xA + yB – – – – – A—A—B—A—B—B—A—A—A—B—

the A:B ratio and the degree of randomness of copolymer formed depends on the quantity and reactivity of monomers and also on the mechanism of the reaction. The properties of the copolymer formed depend on the copolymerization mechanism. There are various copolymerization mechanisms, among which the synthesis by free radical, ionic and polycondensation polymerization are more common.

4.2 Free Radical Copolymerization

Let us take a simple case in which two monomers M_1 and M_2 along with a free radical initiator are present in the reaction mixture. The initiator on decomposition forms a free radical R,• which can react with both the monomers M_1 and M_2 as follows:

$$R\bullet + M_1 \rightarrow RM_1\bullet \qquad ...(4.1)$$

$$R\bullet + M_2 \rightarrow RM_2\bullet \qquad ...(4.2)$$

Thus, two polymer chains are initiated (represented by reactions 4.1 and 4.2), both of which undergo propagation, reacting with either of the monomers M_1 or M_2, present in the feed (reaction mixture). The following four possibilities of propagation reaction arise, which are named PR (propagation reaction):

I. $RM_1\bullet + M_1 \rightarrow RM_1M_1\bullet \ (PR_{11})$ ...(4.3)

II. $RM_1\bullet + M_2 \rightarrow RM_1M_2\bullet \ (PR_{12})$ ...(4.4)

III. $RM_2\bullet + M_1 \rightarrow RM_2M_1\bullet \ (PR_{21})$ ...(4.5)

IV. $RM_2\bullet + M_2 \rightarrow RM_2M_2\bullet \ (PR_{22})$ ...(4.6)

Among these four reactions PR_{11} and PR_{22} are homopolymerization reactions and PR_{12} and PR_{21} are heteropolymerization reactions. They are based on two assumptions: first, that the reactivity of any growing chain depends on the end free-radical only and does not depend on the number and the type of monomers already attached to the chain. The second assumption is that all the four types of additions take place with equal ease. As a result of the reactions (4.3) to (4.6), the polymer chains ending either with M_1 or M_2• are obtained.

The rates of all the four reactions can be given as follows:

$$M_1 \bullet + M_1 \rightarrow M_1M_1 \bullet \qquad R_{11} = k_{11}[M_1 \bullet][M_1] \qquad ...(4.7)$$

$$M_1 \bullet + M_2 \rightarrow M_1M_2 \bullet \qquad R_{12} = k_{12}[M_1 \bullet][M_2] \qquad ...(4.8)$$

$$M_2 \bullet + M_1 \rightarrow M_2M_1 \bullet \qquad R_{21} = k_{21}[M_2 \bullet][M_1] \qquad ...(4.9)$$

$$M_2 \bullet + M_2 \rightarrow M_2M_2 \bullet \qquad R_{22} = k_{22}[M_2][M_2 \bullet] \qquad ...(4.10)$$

where ks are the corresponding rate constants for all the four propagation reactions represented by (4.3)-(4.6). The propagation products $M_1M_1 \bullet$, $M_2M_1 \bullet$, $M_1M_2 \bullet$ and $M_2M_2 \bullet$ symbolize macromolecules with $M_1 \bullet$ and $M_2 \bullet$ terminal groups. In deriving these equations, it is assumed that the reactivity of any growing chain depends only on the free-radical terminal unit and not on the length and type of monomer units attached to the chain. The rate of change in monomer concentration (monomer feed), i.e. the rate of consumption of the monomers during propagation can be given as:

$$-\frac{d}{dt}[M_1] = k_{11}[M_1^\bullet][M_1] + k_{22}[M_2^\bullet][M_2] \qquad ...(4.11)$$

and $$-\frac{d}{dt}[M_2] = k_{12}[M_1^\bullet][M_2] + k_{22}[M_2^\bullet][M_2] \qquad ...(4.12)$$

It has been observed that the number of growing chains of copolymers remains constant during copolymerization, i.e. the concentration of $[M_1 \bullet]$ and $[M_2 \bullet]$ is constant, the rate of conversion of $M_1 \bullet$ and $M_2 \bullet$ is constant and the rate of conversion of $M_1 \bullet$ to $M_2 \bullet$ and $M_2 \bullet$ to $M_1 \bullet$ also remains equal.

$$k_{12}[M_1^\bullet][M_2] = k_{21}[M_2^\bullet][M_1] \qquad ...(4.13)$$

and $$[M_1]/[M_2] = k_{12}[M_1^\bullet]/k_{21}[M_2^\bullet] \qquad ...(4.14)$$

Solving for $M_1 \bullet$ gives

$$[M_1^\bullet] = \frac{k_{21}[M_2^\bullet][M_1]}{k_{12}[M_2]} \qquad ...(4.15)$$

The ratio of disappearance of monomers $[M_1]/[M_2]$ is given in eq. (4.16), assuming a steady state where the rate of a particular chain-end disappearance is equal to the rate of formation of the same chain-end

$$\frac{d[M_1]}{d[M_2]} = \frac{k_{11}[M_1^\bullet][M_1] + k_{21}[M_2^\bullet][M_1]}{k_{22}[M_2^\bullet][M_2] + k_{12}[M_1^\bullet][M_2]} = \frac{[M_1](k_{11}[M_1^\bullet] + k_{21}[M_2^\bullet])}{[M_2](k_{22}[M_2^\bullet] + k_{12}[M_1^\bullet])} \quad ...(4.16)$$

Substitute the value of $M_1\bullet$ from eq. (4.15) to (4.16) gives eq. (4.17)

$$\frac{d[M_1]}{d[M_2]} = \frac{[M_1]}{[M_2]}\left\{\frac{k_{11}k_{21}[M_2^\bullet][M_1]/k_{21}[M_2]) + k_{21}[M_2^\bullet]}{k_{12}k_{21}[M_2^\bullet][M_1]/k_{21}[M_2]) + k_{22}[M_2^\bullet]}\right\} \quad ...(4.17)$$

Divide eq. (4.17) by k_{12} gives eq. (4.18)

$$\frac{d[M_1]}{d[M_2]} = \frac{[M_1]}{[M_2]}\left\{\frac{k_{11}[M_2^\bullet][M_1]/k_{12}[M_2]) + [M_2^\bullet]}{([M_2^\bullet][M_1]/[M_2]) + k_{22}[M_2^\bullet]/k_{21}}\right\} \quad ...(4.18)$$

Substitute k_{11}/k_{12} with r_1, the reactivity ratio of monomer M_1 and k_{22}/k_{21} with r_2, the reactivity ratio of monomer M_2 and cancelling $M_2^\bullet$ gives eq. (4.19)

$$\frac{dM_1}{dM_2} = \frac{[M_1]}{[M_2]}\left\{\frac{(r_{1[M_1]/[M_2]}) + 1}{([M_1]/[M_2]) + r_2}\right\} \quad ...(4.19)$$

Multiplying by $[M_2]$ gives the 'copolymerization equation' (4.20)

$$\text{Copolymer composition} = n = \frac{d[M_1]}{d[M_2]} = \frac{[M_1](r_1[M_1] + [M_2])}{[M_2]([M_1] + r_2[M_2])} \quad ...(4.20)$$

It is difficult to calculate the concentration of free radical in the reaction mixture but eq. (4.20) does not require free-radical concentration to give the copolymer composition. This equation also gives the changing composition of the growing copolymer with the changing monomer feed.

Multiplying eq. (4.20) by $[M_1]$ and $[M_2]$ we get

$$n = \frac{d[M_1]}{d[M_2]} = \frac{[M_1]r_1[M_1] + [M_1][M_2]}{[M_2][M_1] + [M_2]r_2[M_2]} \quad ...(4.21)$$

We can rewrite the equation (4.21) in the following form:

$$\frac{[m_1]}{[m_2]} = \frac{[M_1]}{[M_2]}\left\{\frac{r_1[M_1]+[M_2]}{r_2[M_2]+[M_1]}\right\} \qquad ...(4.22)$$

where $[m_1]$ and $[m_2]$ are the molar concentration of the monomers M_1 and M_2 and m_1/m_2 gives their ratio entered in the copolymer formed.

On dividing eq. (4.22) by $[M_1]$ and $[M_2]$, we get another form of the 'copolymerization equation'.

$$n = \frac{d[M_1]}{d[M_2]} = \frac{r_1([M_1]/[M_2])+1}{r_2([M_2]/[M_1])+1} = \frac{r_1 x+1}{r_2 / x+1} \qquad ...(4.23)$$

The instantaneous composition of monomer feed and copolymer formed is calculated as follows:

$x = m_1/m_2$ is the ratio of monomers in the monomer feed. This eq. (4.23) is also another form of 'copolymerization equation', which gives composition of copolymer 'n' in terms of composition ratio of feed 'x' and reactivity ratios. Since $n = rx$ (Wall equation), a graph is plotted between 'n' and 'x' and the slope of the line gives 'r'. We can represent the monomer concentration in terms of mole fraction also.

(**Note:** In some of the texts 'monomer feed' in terms of mole fraction is represented by 'N' and in some texts by 'f'. The monomer entered the copolymer chain is known as 'copolymer formed' and is represented in some of the texts by 'n' and while in other texts by 'F'. The molar concentration of 'monomer feed' is represented by 'M' and that of the 'copolymer formed' by 'm'.)

In terms of mole fraction, the concentration of monomer M_1 in 'copolymer formed' is represented by n_1 and that of monomer M_2 by n_2. Hence,

$$n_1 = 1 - n_2 = \frac{[m_1]}{[m_1]+[m_2]} \qquad ...(4.24)$$

and

$$n_2 = \frac{[m_2]}{[m_1]+[m_2]} \qquad ...(4.25)$$

There is another important term known as the 'monomer feed', which is the amount of monomers left in the reaction mixture at a particular instant of time. The concentration of monomer feed changes with the progress of the reaction. Monomer feed in terms of mole fraction is represented by N_1 and N_2 for monomers M_1 and M_2, respectively. The concentration of monomer feed is an important parameter for monitoring the synthesis of a copolymer. Monomer feed is represented by the equation (4.26):

$$N_1 = 1 - N_2 = \frac{M_2}{[M_1 + [M_2]]} \qquad ...(4.26)$$

From all the above equations, the copolymer equation can also be given by eq. (4.27)

$$n_1 = \frac{r_1(N_1)^2 + N_1N_2}{r_1(N_1)^2 + 2N_1N_2 + r_2(N_2)^2} \qquad ...(4.27)$$

With the help of eq. (4.27), one can know the instant composition of the copolymer formed by the reactivity ratios; r_1 and r_2 and the monomer feed ratios N_1 and N_2.

4.3 Ionic Copolymerization

There are only a few monomers which form copolymer through ionic copolymerization as compared to the free-radical copolymerization. This is because of the influence of substituents on the ionic charges, which varies the reactivity of the monomers to a great extent. Hence, one monomer becomes more active than the other. As a result, only one monomer enters into polymerization yielding a homopolymer. Thus, best copolymers are obtained only when both the monomers have similar structures, which is why their reactivity is comparable.

In copolymerization, the monomers are capable of forming relatively stable anions. Hence, their anionic initiation is common. However, the cationic copolymerization is also not rare. The presence of substituent in the monomer is the deciding factor for the reactivity of the monomer.

A monomer forms a carbenium ion (carbonium ion is the term in the old literature) to copolymerize through cationic copolymerization. The substituents that increase the electron availability on the double bond increase the reactivity of the monomer towards cationic copolymerization too. The other factor influencing the reactivity is the resonance stabilization of the resulting carbenium ion. As mentioned earlier, there is lesser number of monomers available that have close reactivity values to form good copolymers.

In anionic copolymerization, the reactivity of the carbanion formed is affected by the presence of electron withdrawing substituent in the monomer.

The reactivity ratios of ionic copolymerization are affected by the solvent and the gegen ion in addition to the monomer pair and temperature.

By applying the copolymerization composition equation, the ionic copolymerization can also be understood. The values of the reactivity ratios of monomers are quite different for cationic and anionic copolymerization, which in turn are different from the reactivity ratio values of free radical copolymerization. The reactivity ratios of a particular monomer pair show different values of reactivity ratio if it undergoes cationic and anionic type of copolymerization. If same monomers feed undergoes copolymerization by all the three methods using different initiators, then the composition of the copolymer formed is different. This shows the difference in the mechanisms in all the three copolymerization cases. This difference is a consequence of difference in the behaviour of monomers towards the initiators.

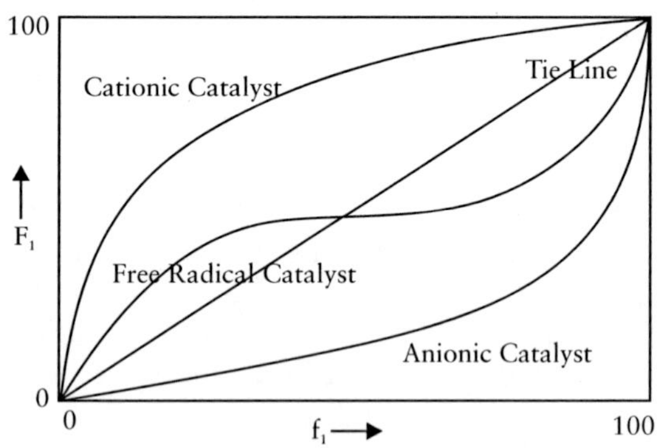

Fig. 4.2: Instantaneous A% composition of copolymer as a function of monomer A% composition showing vast differences in the dependence of F_1 on f_1 when same monomer pair is catalyzed by cationic catalyst, free-radical catalyst and anionic catalyst

We take the example of two monomers A and B whose copolymerization is cationic catalyzed, radical catalyzed and anionic catalyzed. Following is the schematic diagram between monomer A in feed versus A in polymer for the above-mentioned reaction.

4.4 Composition of Copolymers

The values of r_1 and r_2 give a clear picture of copolymer composition and the type, where $r_1 = k_{11}/k_{12}$ and $r_2 = k_{22}/k_{21}$. The reactivity ratios r_1 and r_2 show whether a growing chain carrying the terminal free radical would prefer to add monomer of its own kind or the other monomer. For example, if in a system the value of $r_1 = 0.50$, i.e. in the ratio k_{11}/k_{12}, k_{12} is double the value of k_{11} and $r_2 = 0.50$, i.e. in the ratio k_{22}/k_{21}, k_{21} is also double the value of k_{22}. These values of reactivity ratios show that both the monomer radicals react at twice faster rate when the monomer of another kind is added than when the monomer of its own kind is added. The chemical composition of the copolymer formed depends on the reactivity ratio of the monomers taken for synthesis in addition to the concentration of both the monomers present at a particular instance of time. However, the rate constants for initiation and termination reactions do not appear in the copolymerization equations. Hence, the composition of the copolymer is neither affected by overall rate of the reaction and nor by concentration of the initiator. Reactivity ratio values are also affected by the solvent taken, the inhibitor and the chain transfer agent added during the copolymerization reaction. Chain transfer agent has a pronounced effect on the molecular weight and molecular weight distribution of the copolymer formed.

The values of reactivity ratios can be determined by analyzing the change in feed composition during the early stages of polymerization.

Let us understand the important results obtained from r_1 and r_2 as the behaviour of copolymerization changes with the change in the values of r_1 and r_2. From the point of view of the reactivity ratio, there are two types of copolymerization at two extremes, viz. ideal copolymerization and alternating copolymerization.

1. Ideal Copolymerization: When $r_1 = r_2 = 1$, i.e. $r_1/r_2 = 1$, it gives an idea that the propagation rate constants for propagation reactions PR_{11} and PR_{22} as well as PR_{21} and PR_{22} are same, and all the four reactions take place at the same rate. The probability of adding monomers M_1 and M_2 to M_1• and M_2• depends on the concentration of the monomer. In this case, the monomeric components will have the same ratio in the copolymer formed and the monomer feed throughout the reaction. This is called an 'ideal copolymerization'. The term ideal copolymerization does not mean a desirable type of copolymerization. The copolymer composition can be controlled by initial monomer feed ratio. This copolymerization is also known as 'azeotropic copolymerization'.

2. Alternating Copolymerization: When $r_1 = r_2 = 0$ reactions PR_{11} and PR_{22} do not take place and an alternating copolymer is produced irrespective of the composition of the monomer feed. The copolymer equation for this copolymerization reduces to $d[M_1]/d[M_2] = 1$. This value is a measure of the extent of randomness in copolymer. It is interesting to note here that these monomers are unable to form homopolymers.

But actually, most of the copolymerization reactions lie between ideal and alternating copolymerization where it follows $0 < r_1r_2 < 1$ or both r_1 and $r_2 > 1$.

3. When $r_1 < 1$ and $r_2 < 1$ the PR_{12} and PR_{21} will be preferred to PR_{11} and PR_{22}, which will depend on actual numerical value of r_1 and r_2 and monomer feed ratio.

If $r_1 = r_2 < 1$, the copolymer formed will be richer in M_1 up to $N_1 = 0.5$ and then copolymer formed will be richer in M_2. During the course of the reaction, when N_1 (monomer feed) reaches a value of 0.5, n_1 (copolymer formed) will also be 0.5 but the arrangement of M_1 and M_2 will be random in the copolymer formed. This situation is known as 'azeotropic composition'.

If $r_1 < r_2$, then the 'azeotropic composition' decreases from an N_1 value of 0.5, while if $r_1 > r_2$ then N_1 value increases from 0.5.

In the case when $r_1 < 1$ and $r_2 < 1$, then whether it is $r_1 = r_2$ or $r_1 < r_2$ or $r_1 > r_2$, the copolymer formed in all the cases will be richer in M_1 below 'azeotropic composition' and richer in M_2 above it.

4. If the value of $r_1 > r_2$ then M has a tendency to react with itself and form homopolymer or block copolymer in M_1. Although, if the value of $r_1 < 1$ then M_1 ending chain reacts with another monomer, i.e. with M_2. When $r_1 > 1$ and $r_2 > 1$, the system does not show random copolymerization but either form two homopolymers or under favourable conditions form the block copolymers with long sequences of both the monomers.

When $r_1 < 1$ and $r_2 > 1$, the opposite of the above conditions will take place, i.e. the copolymer formed will be richer in M_2.

5. When the ratio r_1/r_2 is high, i.e. $r_1 > 1$ and $r_2 < 1$, then PR_{11} and PR_{21} will be the preferable reactions than PR_{12} and PR_{22}. This implies that monomer M_1 is consumed preferably and the copolymer formed will be richer in M_1. Finally, monomer M_1 is completely consumed and monomer 2 is left behind in the reaction mixture after which the homopolymer of monomer 2 is formed. If $r_1 >> r_2$ then M_2 would face great difficulty to enter into the copolymer. By keeping the monomer feed ratio $[M_1]/[M_2] < 1$ during the entire course of reaction some M_2 can be incorporated in the copolymer.

6. When $r_1r_2 = 0$ alternate copolymer is produced.

Copolymerization with various reactivity ratios are shown in the following schematic diagram.

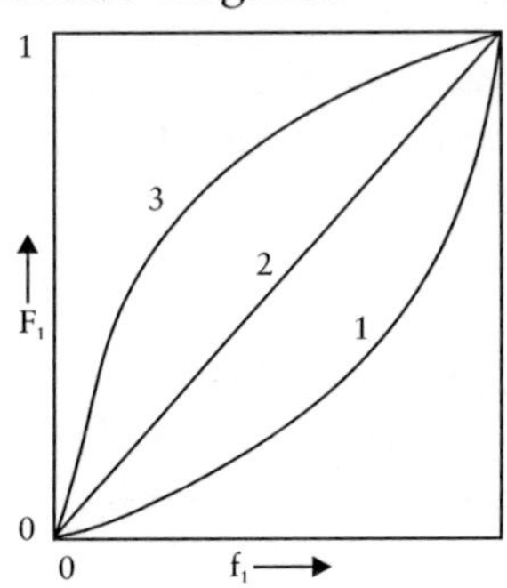

Fig. 4.3: Polymer composition F_1 versus f_1 for ideal copolymerization. Curve 1: low r_1, curve 2: $r_1 = r_2$, curve 3: high r_1

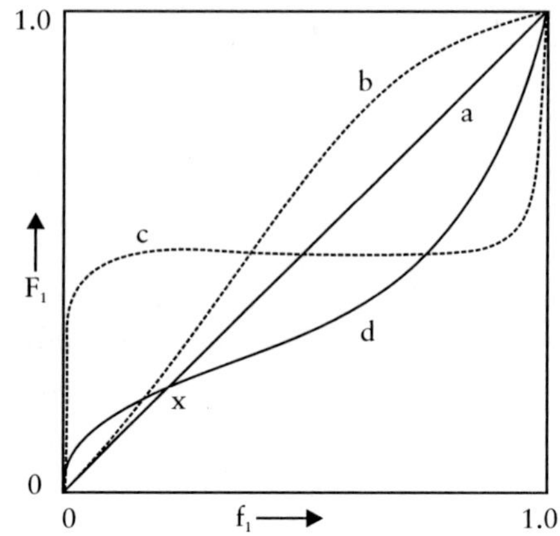

Fig. 4.4: Schematic diagram showing variations of F_1 with f_1 for different cases of random copolymerization

Explanation of the Fig. 4.4

Curve a. The ideal behaviour in copolymerization, i.e. when $r_1 r_2$ =1 approximately

Curve b. The copolymerization is almost ideal.

Curve c. Indicates the alternating system.

Curve d. Radical copolymerization. The curve d crosses the perfectly random case line at the point x is referred to as azeotropic.

4.4.1 Composition Drift

When copolymerization takes place, the ease of entering of both the monomers into the growing chain is different which is because of the difference in the reactivity ratios of both the monomers. The progress of the copolymerization reaction changes the number of monomers in the 'monomer feed'. This change is known as the 'copolymer composition drift'. It has a pronounced effect on the composition of the 'copolymer formed' and as its consequence; the broadening of the distribution of copolymer composition takes place.

In view of the 'copolymer composition drift', one has to see

(i) How prediction of the copolymer composition can be fixed.

(ii) How to predict the behaviour of monomers, which are going to enter into copolymerization.

For predicting the copolymer composition and keeping the composition broadening minimum, one should know the relative reactivity of both the monomers.

For the predictive ability and to have the desired copolymer, one should have Q-e data for the monomers being used.

4.4.2 The Q-e Scheme

This is a semi-empirical method for predicting r_1 and r_2 values when copolymerization takes place by a free radical mechanism. Due to the complexity of copolymerization, the exact prediction of reactivity ratios becomes difficult. The Q-e scheme provides a series of the relativities and polarities of monomers. The basic equation for rate constant is

$$k_{12} = P_1 Q_2 \exp(-e_1 e_2) \qquad \text{...(4.28)}$$

Where the factor P_1 = the factor which characterizes the reactivity of polymeric radical M_1• at the growing chain,

Q_2 = the resonance stability of monomer M_2,

e_1 and e_2 = the terms are proportional to the electrostatic interaction of the permanent charges on the substituent, i.e. the polarization characteristics of monomers M_1 and M_2 and of polymeric radicals M_1• and M_2•, respectively.

This equation can be extended to other propagation reactions (4.3-4.6). We know that $r_1 = k_{11}/k_{12}$ and $r_2 = k_{22}/k_{21}$.

This can also be written as:

$$r_1 = \frac{Q_1}{Q_2}\exp[-e_1(e_1 - e_2)] \qquad \text{...(4.29)}$$

$$r_2 = \frac{Q_2}{Q_1}\exp[-e_1(e_2 - e_1)] \qquad \text{...(4.30)}$$

And $r_1 r_2 = \exp[-(e_1 - e_2)^2]$ ...(4.31)

It is assumed that the same value of e can be applied to monomer and the radical. Styrene is an aromatic compound that copolymerizes with most of the monomers. It is chosen as the standard. The r_1 and r_2 values for a reaction of different monomers

(M_1) are computed with styrene (M_2). Then the standard value of $Q = 1$ and $e = 8$ are assigned for styrene. The values of r_1 and r_2 are determined experimentally. These values are put in the equations (4.29) and (4.30) to get the Q and e values for various monomers. Once Q-e values for various monomers are known, the r_1 and r_2 values for any pair of monomers can be computed. Some of the important conclusions can be drawn from Q-e scheme:

1. Monomers with much difference in their Q values do not copolymerize together.
2. Monomers with a large difference in e values form alternating copolymers.
3. Monomers with close Q and e values show Azeotropic copolymerization.

The monomers which have resonance stabilization have $Q > 0.5$ and in the absence of this stabilization, this value decreases. Hence, Q value is a measure of resonance stabilization.

Similarly, e values are more with electron withdrawing group in the side chain of the monomer. If one monomer has a high e value and the other a low e value, their copolymerization takes place readily.

The major drawback of Q-e scheme is that it does not consider the steric effects.

4.5 Block Copolymer

They do not form naturally but are synthesized.

Butyl rubber is one of the important industrial products synthesized by ionic copolymerization.

Step Copolymer

This is used to synthesize polymers like polyamides and polyesters to have some unique properties in the end product. One such example is Nylone 6,6 and Nylone 6,10. Nylone 6,10 is more flexible and water resistant than Nylone 6,6. This property is caused by the longer carbon chain in Nylone 6,10.

$$H_2N(CH_2)_6NH_2 + HOOC(CH_2)_4COOH$$

Hexamethylene diamine — Adipic acid

$$\rightarrow —NH(CH_2)_6NHCO(CH_2)_4CO— + 2H_2O$$

Nylon 6,6

$$H_2N(CH_2)_6NH_2 + HOOC(CH_2)_8COOH$$

Hexamethylene diamine — Sebacic acid

$$\rightarrow —NH(CH_2)_6NHCO(CH_2)_6CO— + 2H_2O$$

Nylon 6,10

If a long sequence of both the monomers takes place in copolymerization block and graft copolymers result. However, it can be explained that the long sequences of monomers are unlikely in random copolymers. But the increasing importance of these block and graft copolymers led many to develop new techniques for their commercial production.

The long sequence of two monomers can take place in two ways:

1. In block copolymers, the monomers arrange themselves in blocks one after another in the main chain. they can be of many kinds

   —AAAA—BBBBB—AAAA—BBBBB—AB block

   ABA or BAB blocks

   Alternating blocks

   —AAA—BBB—AAA—BBB—

2. In graft copolymers, blocks of one of the monomers attach themselves in a regular fashion to a chain of the other monomer.

   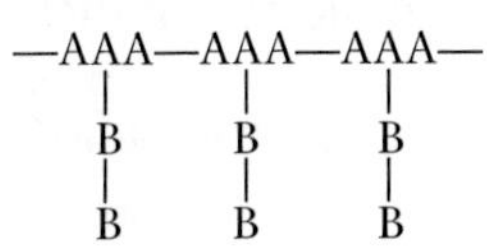

5 Kinetics of Polymerization

Kinetics gives us how various factors influence the rate of reaction. In addition to the rate of reaction, the kinetic study of polymers gives an idea of molecular weight, molecular weight distribution, chemical composition and the extent of polymer conversion also.

5.1 Kinetics of Polycondensation

Step polymerization proceeds by a relatively slow increase in molecular weight of the polymer. The kinetics of polymerization is of prime interest from two viewpoints. The practical synthesis of high polymers requires knowledge of the kinetics of the polymerization reaction. From the theoretical viewpoint, the significant differences between step and chain polymerizations reside in large part in their respective kinetic features.

5.1.1 Kinetics and Statistics of Linear Step Polymerization

The molecules of various chain lengths present in the reaction mixture make the kinetics of step polymerization complicated. As an example, we take the polyesterification reaction. This reaction takes place in the presence as well as in the absence of an acid catalyst. In the absence of catalyst one molecule of dicarboxylic acid acts as auto-catalyst in addition to the one molecule of dicarboxylic acid undergoing esterification. The kinetic study of the reaction becomes easy by using the assumption that the reactivity of functional groups is independent of molecular size. This assumption is known as the principle of equal reactivities of functional groups. This principle makes the kinetic study of step polymerization easy. A single rate constant can be assigned to all the propagation reaction steps. Let us first understand this principle.

5.1.2 Principle of Equal Reactivity of Functional Groups

The kinetic study of step polymerization is difficult to analyze as there are innumerable reactions during propagation. However, the kinetic analysis can be greatly simplified if one assumes that the reactivity of the functional groups of a bifunctional monomer are the same, independent of the size of the molecule to which it is attached. This assumption is referred to as the concept of equal reactivity of functional groups proposed by Flory. According to this concept, the kinetics of a polyesterification reaction of a diol and dicarboxylic acid is the same as that for the esterification of acetic acid and ethanol.

The chemical reactions occur as a consequence of the collision between mutually reactive functional groups. If the collision is successful, then the reaction takes place otherwise either they diffuse or collide again. The reactivity of functional group depends on the frequency of collisions of the group rather than the size of the molecule. As the size of the molecules increase, the rate of molecular diffusion decreases and the functional group collides many more numbers of times thus increasing the probability of reaction. This effect is compensated by the greater duration of each collision giving rise to a larger number of collisions per encounter of the functional group. Hence, the reactivity of a functional group can be expected to be approximately independent of the molecular size.

According to the functionality principle, each of the reacting molecules that take part in polymerization must be bifunctional and give the minimum average functionality of 2 to the system. The average functionality of a reaction between the different kinds of functional groups of the molecules is calculated on the basis of stoichiometric equivalence of both the reacting functional groups. Thus, for reaction between a dicarboxylic acid and a trihydric alcohol

$$R_1 — (COOH)_2 + R_2 — (OH)_3$$

the diacid and the triol should be used in 3 : 2 molar ratio

3 moles of diacid + 2 moles of trihydric alcohol

$3 \times 2 = 6$ and $2 \times 3 = 6$

$6 + 6 = 12$ $3 : 2 = 5$

So 12/5 = 2.4 is the average functionality for the system, which is less than the arithmetic average (5/2 = 2.5) for the system.

Similarly for a reaction between monohydric alcohol and a triacid

$$R_1 — (COOH)_3 + R_2 — OH$$

1 mole + 3 mole

$1 \times 3 + 3 \times 1 = 6$

$3 : 1 = 4$

Average functionality is 6/4 = 1.5, which is less than 2.

Thus, on the basis of above stoichiometric equivalence of reactive functional groups as in the example, a system with a polyfunctional compound (functionality of 3 or more) of a given kind of functional group (say —COOH) and a related monofunctional compound bearing a different kind of functional group (say —OH), would give an average functionality less than 2. So, for any system, if one type of reacting molecules is monofunctional then the system would fail to yield polymer molecules. A double bond (C=C) bearing two π electrons would be viewed as bifunctional, and in fact, on thermal or photoactivation, the π bond forms to yield a biradical intermediate.

With the help of the principle of equal reactivity of functional groups, the kinetics of stepwise polymerization becomes simple and a single rate constant can be assigned to all the propagation steps.

5.1.3 Kinetics of Uncatalyzed Reaction

Most step polymerizations involve bimolecular reactions which are catalyzed. A catalyst is either added externally or they are auto-catalyzed by the reacting acid. The elimination product formed during polycondensation is neglected.

Consider a reaction between two reactants to form a polymer. Each reactant has two terminal functional groups.

$$\begin{aligned} &X-A-X+Y-B-Y+X-A-X(Cat) \\ &\qquad \rightarrow X-A-Y-B+A-X(Cat) \end{aligned} \qquad ...(5.1)$$

The diacid and diol reaction can be written as:

$$\begin{aligned} &\sim COOH + HO \sim + \sim COOH(Cat) \\ &\qquad \rightarrow \sim COOH \sim + H_2O + \sim COOH(Cat) \end{aligned} \qquad ...(5.1a)$$

In auto-catalyzed reaction, the second molecule of acid acts as a catalyst.

The overall rate of reaction can be understood as the rate of decrease of the concentration of reactants. At equilibrium

$$\text{Rate of reaction} = -\frac{d[A]}{dt} = -\frac{d[B]}{dt} = k[A][B][A] \qquad ...(5.2)$$

$X-A-X = [A]$ and $Y-B-Y = [B]$

k = the rate constant for the reaction.

Let's assume that equimolar concentrations of both the monomers are taken in the reaction. The initial concentrations taken are C_0. After time t the concentration of monomers left is C.

Then $[A] = [B] = C$

$$\frac{d[C]}{dt} = kC^3 \qquad ...(5.3)$$

$$\text{Or} \quad \frac{d[C]}{C^3} = kdt \qquad ...(5.4)$$

Integrating the equation (5.4) at $t = 0$, $C = C_0$

$$2kt = \frac{1}{C^2} - \frac{1}{C_0^2} \qquad ...(5.5)$$

Each molecule has two functional groups. If functional groups are represented by n_0 then $n_0 = 2C_0$.

It is convenient to express this result in terms of the extent of reaction p. The extent of reaction is defined as the fraction of the functional group that has reacted in time t. Hence, $(1 - p)$ is the concentration of functional group that remains after time t.

Substituting the concentration terms with the extent of reaction p in the equation (5.5)

$$2ktC_0^2 = \frac{2}{(1-p)^2} - 1 \qquad ...(5.6)$$

A plot of versus t is a straight line. The slope of the curve gives the value of 'k', the rate constant for the polymerization reaction.

5.1.4 Kinetics of Catalyzed Reaction

If a strong acid is used as a catalyst in the polyesterification reaction, it takes up a pseudo second order kinetics. The reaction can be represented as follows:

$$X-A-X+Y-B-Y+\text{catalyst} \rightarrow X-A-B-Y+\text{catalyst} \qquad ...(5.7)$$

The diacid and diol reaction can be written as:

$$\sim COOH + HO \sim + \text{Catalyst} \rightarrow \sim COO \sim + H_2O + \text{Catalyst} \qquad ...(5.7a)$$

$$\text{Rate of reaction} = -\frac{d[A]}{dt} = -\frac{d[B]}{dt} = k'[A][B][\text{Cat}] \qquad ...(5.8)$$

$X - A - X = [A]$ and $Y - B - Y = [B]$

k' is the rate constant for the reaction.

Since the concentration of a true catalyst does not change as reaction proceeds, thus, the concentration of the catalyst can be taken to be constant.

So, $k = k'$ [Catalyst] and equation (5.8) becomes

$$\frac{d[A]}{dt} = k[A][B] \qquad ...(5.9)$$

Let's assume that equimolar concentrations of both the monomers are taken in the reaction. The initial concentrations of the monomers taken are C_0. After time t, the concentration of the monomers remaining unreacted is C.

Then $[A] = [B] = C$

$$\frac{d[C]}{dt} = kC^2 \quad ...(5.10)$$

or $$\frac{d[C]}{c^2} = kdt \quad ...(5.11)$$

Integrating equation (5.11) at $t = 0$, $C = C_0$

$$\left[\frac{1}{C^2}\right]_{C_0}^{C} = [kt]_0^t + \text{Constant}$$

$$kt = \frac{1}{C} - \frac{1}{C_0} \quad ...(5.12)$$

In terms of the extent of reaction, the equation (5.12) takes the form

$$C_0 kt = \frac{1}{(1-p)} - 1 \quad ...(5.13)$$

A plot of $\frac{1}{(1-p)}$ versus t, is a straight line.

In equations (5.6) and (5.13), it has been assumed that the reverse reaction is negligible.

5.1.5 Statistics of Linear Step Polymerization

The statistical analysis of step polymerization gives better and easy results than those from kinetic considerations. The probability of finding a functional group that remains unreacted is $(1 - p)$. We have to find the probability of an x-mer that has $(x - 1)$ reacted functional groups and one unreacted functional group at the end of the molecule. Since the probability of finding one reacted group in the molecule is p, hence, the probability of finding the $x - 1$ functional group in the same molecule is p^{x-1} and that of the unreacted group is $(1 - p)$. Therefore, the probability of finding complete molecule is $p^{x-1}(1 - p)$. This probability is equal to the fraction of all the molecules that are x-mers. For N molecules, the total number of x-mers is given by

$$N_x = Np^{x-1}(1 - p) \quad ...(5.14)$$

If total number of units present is N_0, where

$$N = N_0(1 - p) \quad ...(5.15)$$

So,

$$N_x = N_0(1 - p)^2 p^{x-1} \quad \text{...(5.16)}$$

Equation (5.16) gives the number distribution function for a linear step polymerization at the extent of reaction p.

Similarly, the weight fraction W_x of x-mer is given by

$$W_x = \frac{xN_x}{N_0} \quad \text{...(5.17)}$$

$$W_x = x(1-p)^2 p^{x-1} \quad \text{...(5.18)}$$

If a graph is plotted between N_x and x (the mole fraction based on the number), then according to equation (5.16), the following trend is observed.

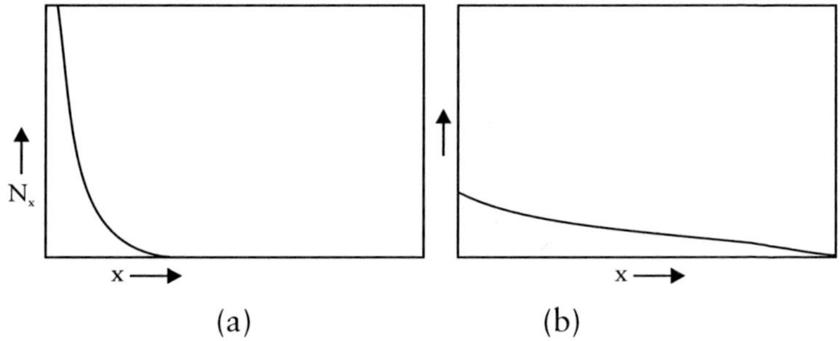

Fig. 5.2: Mole fraction distribution of chain molecules in a linear condensation polymer for extents of reaction (a) 0.095 (b) 0.099

This graph shows that low molecular weight species will be present in more quantity in the reaction mixture than the high molecular weight species at low as well as high extents of reaction.

If weight fraction is taken into consideration and a graph is plotted between W_x and x according to the equation (5.18), it is observed that low molecular weight species will be present in a negligible amount as compared to high molecular weight species.

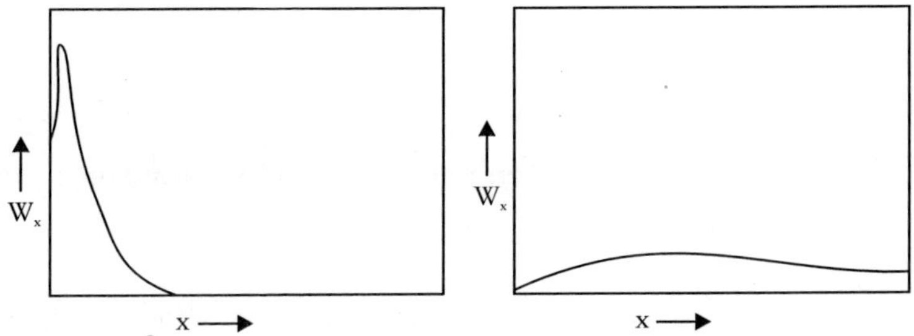

(a) (b)

Fig. 5.3: Weight fraction distribution of chain molecules in linear condensation for extents of reaction (a) 0.095 (b) 0.099

Degree of Polymerization $(\overline{DP})$

This is given by the number of molecules initially present in the reaction mixture, divided by the number of molecules left unreacted after time t.

$$(\overline{DP}) = \frac{N_0}{N} = \frac{N_0}{N_0(1-p)} = \frac{1}{(1-p)} \qquad ...(5.19)$$

In order to get high molecular weight polymer, the extent of reaction should by high. For example to get $(\overline{DP}) = 50,$ the extent of reaction should be 98%. If $p = 0.98$

$$(\overline{DP}) = \frac{1}{1-0.98} = 50.$$

5.2 Kinetics of Chain Polymerization

We have discussed in Chapter 2 that the mechanism of chain polymerization takes place in three steps, viz. initiation, propagation and termination. There are three types of initiation—free radical, cationic and anionic. Let's discuss the kinetics of all the three.

5.2.1 Kinetics of Radical Polymerization

The overall rate of free radical polymerization is determined as follows:

Suppose the free radical initiator I is dissociated to form two free radicals per molecule

1. $I \xrightarrow{k_d} 2R^{\bullet}$ ...(5.20)

This free radical reacts with monomer M to initiate the chain

2. $R^{\bullet} + M \xrightarrow{k_i} RM^{\bullet}$ ...(5.21)

Propagation

3. $RM^{\bullet} + M \xrightarrow{k_p} RMM^{\bullet}$ ...(5.22a)

$RMM^{\bullet} + M \xrightarrow{k_p} RMMM^{\bullet}$ ...(5.22b)

These propagation steps can be written in general as

$$RM^{\bullet}_{(n-1)} + M \xrightarrow{k_p} RM^{\bullet}_n \qquad ...(5.23)$$

4. Termination

It can take place in two ways

(a) Combination

$$M^{\bullet}_n + M^{\bullet}_m \xrightarrow{k_{tc}} M_{n+m} \qquad ...(5.24a)$$

(b) Disproportionation

$$M^{\bullet}_n + M^{\bullet}_m \xrightarrow{k_{td}} M_n + M_m \qquad ...(5.24b)$$

If we assume that both the termination steps are kinetically equivalent, we can write

$$M^{\bullet}_n + M^{\bullet}_m \xrightarrow{k_t} P \qquad ...(5.24c)$$

where $k_t = k_{tc} + k_{td}$

The rate of initiation can be written as

$$v_i = \left(-\frac{d}{dt}[M]\right) = k_i[R^{\bullet}][M] = 2fk_d[I] \qquad ...(5.25)$$

where v_i = rate of initiation, k_i = initiation rate constant, f = fraction of radicals formed that could initiate the chain, k_d = dissociation constant of the initiator. This is an important unimolecular equation that controls the entire chain polymerization.

(c) The rate of termination is written as (for example termination by combination)

$$v_t = -\left(\frac{d}{dt}[M^{\bullet}]\right) = 2k_{tc}[M^{\bullet}]^2 \qquad ...(5.26)$$

k_{tc} = rate constant for termination by combination

At steady state $v_i = v_t$

So $2fk_d[I] = 2k_{tc}[M^{\bullet}]^2$

$$[M^{\bullet}] = \left(\frac{fk_d[I]}{k_t}\right)^{1/2} \qquad ...(5.27)$$

Rate of propagation which gives the overall rate of reaction

$$v_p = -\frac{d}{dt}[M] = k_p[M][M^{\bullet}] \qquad ...(5.28)$$

in which, substitution of the value of $[M^{\bullet}]$ from equation (5.27) we get

$$v_p = k_p\left(\frac{fk_d[I]}{k_t}\right)^{1/2}[M] \qquad ...(5.29)$$

If $k_p\left(\frac{fk_d}{k_t}\right)^{1/2} = K$ then

$$v_p = K[I]^{1/2}[M] \qquad ...(5.30)$$

Rate of formation of polymer is proportional to the square root of initiator concentration and to the first power of monomer concentration. Equation (5.30) would hold when termination takes place by combination and there is no auto-acceleration reaction.

Kinetic Chain Length

$$v = \frac{\text{rate of propagation}}{\text{rate of initiation}} = \frac{v_p}{v_i} = \frac{v_p}{v_t} \qquad ...(5.31)$$

$$v = \frac{k_p}{2(fk_dk_t)^{1/2}}\frac{M}{I^{1/2}} = \frac{k_p^2[M]^2}{2k_tv_p} \qquad ...(5.32)$$

The kinetic chain length v is related to the degree of polymerization $\overline{X}_n$ by the following equation:

1. for termination by combination $\overline{X}_n = 2v$
2. for termination by disproportionation $\overline{X}_n = v$

This shows that the actual size of the polymer molecule depends on the mechanism of termination, but the kinetic chain length is independent of the termination mechanism.

Degree of Polymerization

$$DP = \frac{R_p}{R_i} = \frac{k_p[M]^{(k_d f[I])^{1/2}} \Big/ (k_t)^{1/2}}{2k_d[I]f} \qquad ...(5.33)$$

$$DP = \frac{R_p}{R_i} = \frac{k_p[M]}{k_t^{1/2}\, 2k_d^{1/2} f^{1/2} I^{1/2}}$$

$$\text{If } \frac{k_p}{(2k_d k_t f)^{1/2}} = k''$$

$$\text{then } DP = \frac{[M]}{[I]^{1/2}}.k'' \qquad ...(5.34)$$

Equation (5.34) shows that the molecular weight of polymer is directly proportional to $[M]$ and $[I]^{1/2}$.

Temperature has a pronounced effect on the rate of reaction since dissociation of becomes fast. But at the same time, the average molecular weight or chain length decreases. There is some temperature limit for every polymerization reaction and no polymerization takes place after this temperature. This temperature is known as the Ceiling temperature.

5.2.2 Kinetics of Cationic Polymerization

In cationic polymerization, the initiator on dissociation liberates a proton, which is able to react with the monomer (which is prone to the attack by a cation) to form an ion pair before initiation step of polymerization. Then this ion pair reacts with another molecule of monomer in initiation step. Then the monomer cation starts reacting with other molecules of monomer and propagation reaction starts. Termination occurs with the loss of positive charge of the cationic chain. Sometimes the positive charge shifts to another species then it is termed as chain transfer. This polymerization is highly influenced by the changes in the reactant concentration, composition of solvent and the reaction conditions. The reaction is too fast hence is carried out at low temperature. But due to high rate of the reaction, it is difficult to measure the rate and special techniques are used to measure the rate.

The schematic mechanism of cationic chain polymerization can be given as follows:

1. Dissociation of initiator: $A + BH \rightleftharpoons H^+B^-A$
2. Initiation: $H^+B^-A + M \xrightarrow{k_i} HM^+B^-A$
3. Propagation: $HM^+B^-A + M \xrightarrow{k_p} HMM^+B^-A$
4. Termination: $HM_nM^+B^-A \xrightarrow{k_t} HM_nM + H^+B^-A$
5. Transfer to monomer:

$$HM_nM^+B^-A + M \xrightarrow{k_{tr}} HM_nM + M^+B^-A$$

The rate of initiation

$$v_i = k_i[H^+][M] \qquad ...(5.35)$$

$[M]$ = concentration of monomer

Rate of termination is

$$v_t = k_t[M^+] \qquad ...(5.36)$$

The cationic termination has first order kinetics.

On applying steady state

$$k_i[H^+][M] = k_t[M^+]$$

$$[M^+] = \left(\frac{k_i}{k_t}\right)[H^+][M] \qquad ...(5.37)$$

Rate of propagation, and putting the value of M^+ from equation (5.37)

$$R_p = -\frac{d}{dt}[M] = k_p[M^+][M] = k_p\frac{k_i}{k_t}[H^+][M]^2 \qquad ...(5.38)$$

The overall polymerization rate is second order with respect to monomer concentration.

Degree of Polymerization $(\overline{DP})$

When termination is the predominating step

$$\overline{DP} = \frac{R_p}{R_t} = \frac{k_p[M][M^+]}{k_t[M^+]} = \frac{k_p[M]}{k_t} \qquad ...(5.39)$$

When transfer is the predominating step

$$\overline{DP} = \frac{R_p}{R_{tr}} = \frac{k_p[M][M^+]}{k_{tr}[M][M^+]} = \frac{k_p}{k_{tr}} \qquad ...(5.40)$$

When termination and transfer both take place

$$\overline{DP} = \frac{R_p}{R_t + R_{tr}} = \frac{k_p[M][M^+]}{k_t[M^+] + k_{tr}[M][M^+]} = \frac{k_p[M]}{k_t + k_{tr}[M]} \qquad ...(5.41)$$

Hence, the degree of polymerization is independent of initiator concentration. When chain transfer is the predominant step, the degree of polymerization becomes independent of monomer concentration too.

Kinetic Chain Length

It is given by

$$\overline{X}_n = \frac{\text{rate of propagation}}{\text{rate of formation of dead chains}} \qquad ...(5.42)$$

$$\overline{X}_n = \frac{k_p[M^+][M]}{k_t[M^+] + k_{tr}[M^+][M]} \qquad ...(5.43)$$

If termination is predominant over transfer reaction, the equation becomes

$$n = \frac{k_P[M^+][M]}{k_T[M]} = \frac{k_p[M]}{k_t} \qquad ...(5.44)$$

and transfer reaction predominates then

$$n = \frac{k_P[M^+][M]}{k_{TR}[M^+][M]} = \frac{k_p}{k_{tr}} \qquad ...(5.45)$$

The Effect of Temperature

The rate of cationic polymerization is inversely proportional to the temperature, which can be explained on the basis of Arrhenius equation as follows:

According to Arrhenius equation, the rate constants are given as:

$$k_i = A_i e^{-E_i/RT} \quad \text{...(5.46a)}$$

$$k_p = A_p e^{-E_p/RT} \quad \text{...(5.46b)}$$

$$k_t = A_t e^{-E_t/RT} \quad \text{...(5.46c)}$$

The overall rate constant k_R is given by

$$k_R = \frac{k_i k_p}{k_t} \quad \text{...(5.47)}$$

in which the values of *k*s are taken from equations (5.46a)-(5.46c) so that the value of total energy obtained is:

$$E = E_i + E_p - E_t \quad \text{...(5.48)}$$

Energy for chain length is represented as:

$$E_x = E_p - E_t \quad \text{...(5.49)}$$

In cationic polymerization, the propagation step needs less energy since it is a combination of cation with a neutral molecule. So E_p is less than E_i, E_t and E_{tr}. When the value of E is negative, the value of K_p increases. Chain length always decreases with the increase in temperature because E_t is always greater than E_p.

5.2.3 Kinetics of Anionic Polymerization

The initiator AB dissociates to A^+ and B^- which remain as an ion pair.

1. Initiator dissociation: $AB \rightleftharpoons A^+ + B^-$ ...(5.50)

   and K is the equilibrium constant for above reaction
2. Initiation: $B^- + M \xrightarrow{k_i} BM^-$ ...(5.51)
3. Propagation: $BM^- + M \xrightarrow{k_p} BMM^-$ ...(5.52)
4. Transfer: $BM_xM^- + HB \xrightarrow{k_{tr}} BMMH + B^-$ ...(5.53)

The termination step does not take place in anionic polymerization, rather a proton transfers from a solvent molecule to polymer chain. To make it clear let's take an example where AB is KNH_2, the initiating agent and HB is NH_3, the solvent. Equation (5.53) takes the form of equation (5.54)

$$NH_2M_n^- + NH_3 \xrightarrow{k_{tr}} NH_2M_nH + NH_2^- \quad \text{...(5.54)}$$

1. Rate of initiation

$$v_i = -\frac{d}{dt}[M_i] = k_i[AB][M] \quad ...(5.55)$$

2. Rate of termination (transfer) is given by

$$v_{tr} = -\frac{d}{dt}[M_{tr}] = k_{tr}[M^-][HB] \quad ...(5.56)$$

Assuming steady state

$$v_i = v_{tr} = k_{tr}[M^-][HB] = k_i[AB][M] \quad ...(5.57)$$

to get the value for $[M^-] = \frac{k_i}{k_{tr}} \frac{[AB][M]^2}{[HB]}$ ...(5.58)

3. Rate of polymerization: $R_p = -\frac{d}{dt}[M] = k_p[M][M^-]$

inserted the value from equation (5.58) in the above equation we get

$$R_p = \frac{k_p k_i}{k_{tr}} \frac{[AB][M]^2}{[HB]} \quad ...(5.59)$$

The rate is of second order with respect to monomer concentration and that of first order with respect to initiator concentration.

The Degree of Polymerization

$$\overline{X}_n = \frac{R_p}{R_{tr}} = \frac{k_p[M]}{k_{tr}[HB]} \quad ...(5.60)$$

Equation (5.60) shows that the degree of polymerization depends upon the solvent as it takes part in the transfer mechanism. The value of k_p/k_{tr} is not high, which demonstrates that the transfer reaction is competitive with propagation step. Hence, in anionic polymerization, polymers of low molecular weight are obtained. Another interesting fact about anionic polymerization is that all the chains grow at the same rate, thus, the polymer obtained is monodisperse.

Part – C
Characterization

6 Molecular Weight of Polymers

Many of the properties of a polymeric material, e.g. solubility, melt viscosity, solution viscosity and mouldability, etc. depend on its molecular weight. Therefore, the knowledge of molecular weight becomes important.

6.1 Determination of Molecular Weight

The accuracy of the molecular weight measurement of polymers is low compared to the low molecular weight compounds. Its accuracy decreases as the extent of polydispersity increases. There are several methods to determine the molecular weight of polymers that are either relative or absolute. The relative method needs calibration of the known samples. The absolute methods give various weight averages, e.g. the methods based on colligative properties yield number average molecular weights ($\bar{M}_n$) and the methods in which size estimation of polymer is done, give weight average molecular weight ($\bar{M}_w$). In addition to this, there are methods in which the chemical functional groups are estimated to find $\bar{M}_n$. There are four types of representations for average molecular weights in case of polymers and therefore, there are several methods to determine them.

Following are some well-known experimental methods for determining molecular weight.

1. Number average molecular weight and molecular weight distribution

(a) Osmotic pressure

(b) Intrinsic viscosity

2. Weight average molecular weight and molecular weight distribution

(a) Ultracentrifuge sedimentation

(b) Diffusion

(c) Light scattering

3. z and z + 1 average molecular weight and molecular weight distribution

(a) Ultracentrifuge sedimentation

Above are primary methods. There are some secondary methods like

(a) Fractional precipitation

(b) Gel permeation chromatography

(c) High-performance liquid chromatography (HPLC)

(d) Electrophoresis

We will discuss some of them.

1. *Determination of Number Average Molecular Weight ($\bar{M}_n$)*

Molecular weight is a dimensionless quantity and is defined as the average mass of the molecule divided by one-twelfth the mass of an atom of the ^{12}C. This quantity is known as molecular mass or molar mass (in SI units) but popularly called molecular weight. The number average molecular weight $\bar{M}_n$ is determined by counting the number of molecules irrespective of shape or size, present in a unit mass of the polymer:

6.2. Carothers Equation

W.H. Carothers did the pioneering work on step polymerization. He developed a simple equation for the calculation of the molar mass of a condensation polymer. He related number average degree of polymerization with respect to monomer units by the following equation:

$$\overline{X}_n = \frac{N_0}{N} \qquad ...(6.1)$$

where N_0 = number of monomer molecules present initially, and N = number of monomer molecules left after time t.

In a step polymerization reaction, it is assumed that there are equal numbers of functional groups attached to two monomers. $\bar{X}_n$ can be related to the extent of reaction p at time t as follows:

$$p = \frac{\text{number of functional groups that have reacted}}{\text{number of functional groups present initially}} \quad ...(6.2)$$

$$p = \frac{N_0 - N}{N_0} \quad ...(6.3)$$

$$p = 1 - \frac{N}{N_0}$$

$$\frac{N_0}{N} = \frac{1}{1-p} \quad ...(6.4)$$

Combining equations (6.1) and (6.4), we have

$$\bar{} \qquad \frac{}{} \quad ...(6.5)$$

This is the well-known Carothers equation. This equation is applicable to step reactions of the type $RA_2 + RB_2$ and $ARB + RA_2$. In these reactions, there is a stoichiometry balance in the number of reactive functional groups. This equation shows very high extent of reaction. In step polymerization, the value of $p \geq 0.99$ is needed in order to have the value of the degree of polymerization to be 100 or more. This condition is fulfilled when the reactants of high purity and high reactivity are taken. The $\overline{X}_n$ can also be related to number average molar Mass $\overline{M}_n$ as follows:

$$\overline{M}_n = M_0 \overline{X}_n \quad ...(6.6)$$

where M_0 is the mean molar mass of a monomer unit of which the value is

$$M_0 = \frac{\text{molar mass of repeat unit}}{\text{number of monomer units in the repeat unit}} \quad ...(6.7)$$

Now we calculate the value of $\overline{X}_n$, the number average degree of polymerization of linear step polymerization. Let's suppose a reaction between RA_2 and RB_2, where RB_2 is taken in excess. There, the ratio of the number of functional groups present when $t = 0$ is termed as reactant ratio 'r'. In this polymerization

$$r = N_A/N_B \qquad ...(6.8)$$

where N_A and N_B are the numbers of functional groups that are present at the commencement of the reaction. Since there are two functional groups per molecule, so

$$N_0 = \frac{N_A + N_B}{2} \qquad ...(6.9)$$

In order to omit N_A from equation (6.9) with the help of equation (6.8) we get

$$N = \frac{N_B(1+r)}{2} \qquad ...(6.10)$$

It is customary to represent the extent of reaction p in terms of the functional group which is present in a lesser amount (A group in the present case). So,

the number of unreacted A groups = $N_A - pN_A = rN_B\ (1 - p)$ ...(6.11)

and the number of unreacted B groups = $N_B - pN_A = N_B\ (1 - rp)$ ...(6.12)

Thus, the total number of unreacted groups is:

$$N = \frac{rN_B(1-p) + N_B(1-rp)}{2} \qquad ...(6.13)$$

or,

$$N = \frac{N_B(1+r-2rp)}{2} \qquad ...(6.14)$$

Substitution of equation (6.10) and equation (6.14) in equation (6.1) gives the general Carothers equation:

$$\overline{X}_n = \frac{(1+r)}{1+r-2rp} \qquad ...(6.15)$$

In this equation, if the value of r is equal to 1, then it'll result in equation (6.5).

Equation (6.15) can also be used when in addition to two bifunctional reactants a monofunctional reactant is also added to the reaction mixture in order to control the degree of polymerization $\bar{X}_n$.

Flory did a statistical analysis of Carother's equation for the prediction of the size distribution of step polymerization. Let's take a reaction mixture in which the reactants RA_2, RB_2 and ARB are in equal stoichiometry. After time t the extent of reaction is p. The probability $P(x)$ of a polymer molecule with x monomers can be calculated at a time t. The number of linkages in this mer is $(x–1)$. Let's suppose, the probability of one functional group is p; the probability of a sequence for two linkages is p^2 and for three linkages it will be p^3. Similarly, for $(x–1)$ linkages the probability would be equal to $p^{(x-1)}$. In this molecule, the probability of the last unit of monomer that contains the unreacted functional group would be $(1–p)$. The total probability of this molecule is:

$$P(x) = (1 - p)\, p^{(x-1)} \qquad ...(6.16)$$

Since $P(x)$ is the probability that the molecule that has x monomers, this also is the mole fraction of x mers. If total number of polymer molecules present at time t is N, then the probability of total number of x-mers $N(x)$ is given by:

$$N(x) = N(1 - p)\, p^{(x-1)} \qquad ...(6.17)$$

N cannot be measured, so, in order to omit N from the equation (6.17) we put

$N = N_0\,(1 - p)$ in equation (6.17) and the value of

$$N(x) = N_0\,(1 - p)^2\, p^{(x-1)} \qquad ...(6.18)$$

This is the number distribution for a linear step polymerization the extent of reaction p.

In the same way, weight fraction W_x of x-mers can be given by

$$W_x = \frac{\text{total mass of mulecules with degree of polymerization } x}{\text{total mass of all the molecules}} \qquad ...(6.19)$$

Thus, neglecting the end groups

$$W_x = \frac{N_x(xM_0)}{N_0M_0} = \frac{xN_x}{N_0} \quad ...(6.20)$$

Combining equations (6.18) and (6.20)

$$W_x = x\,(1-p)^2p^{(x-1)} \quad ...(6.21)$$

Equations (6.18) and (6.21) give the most probable distribution (Shultz-Flory distribution). Important conclusions can be drawn when a graph is plotted between $N(x)$ vs x and W_x vs x for various extents of reaction Fig. 5.2 and Fig. 5.3.

The mole fraction $P(x)$ decreases with the increase in the number of monomer units in the polymer chain. The weight fraction distribution curve shows a maximum which is close to $\bar{X}_n$. As the extent of reaction increases the maxima moves to higher values of x and the weight fraction of monomer becomes smaller gradually.

With the help of distribution functions $P(x)$ and $W(x)$ the molar mass averages can be calculated.

The equation for number average molar mass is

$$\overline{M}_n = \sum P(X)M_x \quad ...(6.22)$$

where $M_x = xM_0$ ...(6.23)

Using equation (6.17), the equation (6.22) becomes

$$\overline{M}_n = \sum XM_0(1-p)p^{(x-1)} \quad ...(6.24)$$

i.e. $$\overline{M}_n = M_0\,(1-p)\sum xp^{(x-1)}$$

The weight average molecular weight is

$$\overline{M}_w = M_0\,\frac{(1+p)}{(1-p)} \quad ...(6.25)$$

The weight average degree of polymerization is

$$\overline{X}_w = \frac{(1+p)}{(1-p)} \quad ...(6.26)$$

The polydispersity index is given by

$$\frac{\overline{X}_w}{\overline{\overline{M}}_n} = (1 + p) \qquad \text{...(6.27)}$$

6.3 End-group Analysis

This method of analysis can only be used when the polymer chain has groups which can be analyzed by chemical methods e.g. if —COOH group is attached to the polymer chain then the analysis can be done by titrating with NaOH. Similarly, the amino group can be titrated with any chosen acid. The sensitivity of the results of analysis decreases with increase in the length of polymer chain and decreases with the decrease in the number of functional groups to be analyzed. The number average molecular weight of polymers with $\overline{M}_n$ value less than 15,000 g mol^{-1} can be analyzed with this method.

6.4 Colligative Property Method

The addition of a solute changes the physical properties of a solvent. This change is a function of the concentration of the amount of solute added. If a physical property is represented by Q, then its concentration dependence can be given by an equation having power series. If the concentration of the solution is represented by N (the number of molecules of solute present in unit volume of the solution), then

$$Q = Q_0 + K_1N + K_2N^2 + K_3N^3 + \cdots\cdots \qquad \text{...(6.28)}$$

where Q_0 = the corresponding physical property of the pure solvent. This is known as the Virial equation and K_1, K_2, K_3, etc. are the Virial constants. The physical properties and the K_1 values are independent of the chemical nature of the solute are useful to determine the molecular weight of the solute. In the view of this the colligative property is defined as the function of the number of solute molecules present in per unit volume of solution and is unaffected by the chemical nature of the solute. Following are the colligative properties where the solvent in the solution is in equilibrium with pure solvent either across a membrane or in another phase, which are: partial pressure of the solvent vapour in equilibrium with the solution, elevation of the boiling point, depression of freezing point and the pressure on the solution which is in osmotic equilibrium with the pure solvent.

$$\Delta p_1 = \frac{p_1^0 V_1^0}{N_0} N + K_2' N^2 + K_3' N^3 + \ldots\ldots \quad \ldots(6.29a)$$

$$\Delta T_f = \frac{RT_1^2 V_1^0}{L_f N_0} N + K_2'' N^2 + K_3'' N^3 + \ldots\ldots \quad \ldots(6.29b)$$

$$\Delta T_b = \frac{RT_b^2 V_1^0}{L_v N_0} N + K_2''' N^2 + K_3''' N^3 + \ldots\ldots \quad \ldots(6.29c)$$

$$\pi = \frac{RT}{N_0} N + K_2'''' N^2 + K_3'''' N^3 + \ldots\ldots \quad \ldots(6.29d)$$

In each case, K_1 is written explicitly. V_1^0 is the molar volume of the solvent at the various temperatures at which the values Δp_1, ΔT_f and ΔT_b are measured. N_0 is the Avogadro number. Δp_1 is the difference between the partial pressure of solvent vapour above the solution and the vapour pressure of the pure solvent $p_1{}^0$ at the same temperature.

ΔT_f is the difference between the f_p of the solution and the pure solvent T_f, of which the latent heat of fusion is L_f.

ΔT_b is the difference between the boiling point of the solution and that of the pure solvent T_b, whose latent heat of vaporization is L_v.

Π is the osmotic pressure of the solution at a temperature T.

For low molecular weight compounds, N^2 and N^3 become negligible as compared to the first term. Hence, the values of Δp_1, ΔT_f, ΔT_b and Π can be measured accurately.

If the size of the molecules of the solute and the solvent are comparable and the solution is dilute, the terms N^2, N^3, etc. become negligible as compared to the first term. As a result, Δp_1, ΔT_f, ΔT_b and Π become directly proportional to the number of solute molecules per unit of solution N. The solution is said to be ideal. If proportionality constants and weight concentration of the solution are known, then the number average molecular Weight $\overline{M}_n$ follows the following relation:

$$N = \frac{N_0 C}{\overline{M}_n}$$

where c = weight concentration of the solute, i.e. the weight of solute per unit volume of the solution.

But in the case of polymers, the quantities Δp_1, ΔT_f, ΔT_b and Π become small and immeasurable. The conditions are set so that

$$\Delta Q = Q - Q_0 = K_1 N + K_2 N^2 \qquad \text{...(6.30)}$$

We know the value of K_1 but not that of K_2, hence, N cannot be determined. The only solution to this problem is to replace N by c. For, if we add total N solute particles having a range of molecular weights M_1, M_2 ....M_i, the total weight becomes c in unit volume of solution. Then

$$N\overline{M}_n = \sum N_i M_i \qquad \text{...(6.31)}$$

and

$$c = \frac{\sum N_i M_i}{N_0} \qquad \text{...(6.32)}$$

Combining equations (6.31) and (6.32) we get

$$N = \frac{N_0 c}{\overline{M}_n} \qquad \text{...(6.33)}$$

which is analogous to the equation for the simple solutes.

From eq. (6.30) and eq. (6.33) we get

$$\Delta Q = \frac{K_1 N_0}{\overline{M}_n} c + K_2 \left(\frac{N_0}{\overline{M}_n} \right)^2 c^2 \qquad \text{...(6.34)}$$

or

$$\frac{\Delta Q}{c} = \frac{K_1 N_0}{\overline{M}_n} + K_2 \left(\frac{N_0}{\overline{M}_n} \right)^2 c \qquad \text{...(6.35)}$$

This is the basic equation for the determination of $\overline{M}_n$. If a graph is plotted for $\Delta Q/c$ against c, then intercept is $K_1 N_0 / \overline{M}_n$.

From here, K_1 can be calculated and subsequently $\overline{M}_n$ is obtained. The experimental data shows that Δp_1, ΔT_f and ΔT_b can help determine the molecular weight of the polymers only of an average molecular weight up to 10^3. But the molecular weight of the order of $5\times10^3 - 10^6$ can be determined by osmotic pressure method.

6.5 Osmotic Pressure Method

If we put $$K_1 = \frac{RT}{N_0} \quad \text{...(6.36)}$$

And $$B = K_2\left(\frac{N_0}{\overline{M}_n}\right)^2 \quad \text{...(6.37)}$$

then, the equation (6.2d) becomes

$$\pi = \left(\frac{RT}{\overline{M}_n}\right)\times c + Bc^2 \quad \text{...(6.38)}$$

and

$$\frac{\pi}{c} = \frac{RT}{\overline{M}_n} + Bc \quad \text{...(6.39)}$$

A graph is plotted between π/c and c. When $c = 0$, the intercept obtained is $RT/\overline{M}_n$. By putting the values for π/c, R and T we get the value of $\overline{M}_n$. The slope of the line gives the value for B. The values of both the terms B and $\overline{M}_n$ give an idea of their contribution to osmotic pressure. The second Virial coefficient depends on several factors. The effect of temperature and solvent on second virial coefficient is connected to the variation of average dimension of polymer molecules. When the attractive and repulsive forces of each segment with other segments and solvent molecules come into equilibrium, the polymer molecules are allowed to perform a free rotation about the bonds between segments. This is called unperturbed configuration. Under this condition, the solvent behaves ideally and the equation for osmotic pressure becomes.

$$\pi = \frac{RT}{\overline{M}_n} c$$

This situation arises at the theta temperature. Above theta temperature, the polymer molecule expands and second virial coefficient becomes large and positive. When this occurs the solvent is described as 'good solvent'. The solvent is known as 'poor solvent' when the polymer molecule contracts and the second virial coefficient approaches zero.

Osmometry can either be Vapour pressure Osmometry or Membrane Osmometry.

6.5.1 *Vapour Pressure Osmometry*

This method is an indirect way to measure the lowering in vapour pressure of a solvent by the addition of a polymeric solute. This lowering in vapour pressure is directly proportional to the number of solute particles. The vaporization of solvent is carried out under isothermal condition from the pure solvent as well as from the solution. The vaporization of solvent from the pure solvent is faster than the vaporization of the solvent from a polymer solution. There arises a small change in temperature that results from different rates of evaporation of solvent from and condensation onto droplets of pure solvent and polymer solution maintained in an atmosphere of solvent vapour. This temperature difference is proportional to the lowering of the vapour pressure of the polymer solution at equilibrium. This lowering is proportional to the number average molecular weight of the polymer. This is a rapid method and needs less amount of polymer sample. Measurements are made at several concentrations and extrapolated to zero concentration. The number average molecular weight of polymers that have a molecular weight less than 30,000 g/mole can be determined by vapour pressure Osmometry.

6.5.2 *Membrane Osmometry*

In the case of high molecular weight polymers, the selection of semi-permeable membrane is easy. The $\overline{M}_n$ of polymers between 30,000-15,00,000 can be determined by this method. The principle of the method states that the chemical potential (μ) of a solvent in a solution is lower than that of the pure solvent when the pure solvent and solution are separated by a semi-permeable membrane. The molecules of the solvent can move into the solution through a semi-permeable membrane until the osmotic pressures of both the compartments become equal.

6.6 Molecular Weight Determination and Effect of Solvent

The molecular weight of a polymer can only be determined in solution phase, but if a polymer is insoluble, its molecular weight cannot be determined. When a polymer is dissolved in a solvent different types of interactions develop between the solvent and the polymer chain. As a result of various interactions, if polymer chain expands when dissolved, the solvent is called 'good solvent'. But, if the polymer chain contracts then that solvent is called 'poor solvent'. There is a third variety of solvents which do not exert any effect on the polymer chain and it acquires unperturbed dimensions. These are called 'theta solvents'. This is an important class of solvents. They are used to calculate various physical properties of polymers. They are most suitable for the determination of molecular weight determination of polymers. It is interesting to note that with the variation of temperature the same solvent can act as a good, poor and theta solvent. The fourth variety of solvents is in which there is phase separation and polymer gets precipitated. They are known as 'non-solvents'. The behaviour of polymer chains in a particular solvent at a particular temperature is measured in terms of end-to-end distance. Flory has defined this behaviour as chain expansion factor α which can be calculated by the following equation:

$$\left(\frac{\overline{r}^2}{\overline{r}_\theta^2}\right) = \alpha \qquad ...(6.40)$$

where $\left(\overline{r}^2\right)^{1/2}$ = average root mean square end-to-end distance

and $\left(\overline{r}_\theta^2\right)^{1/2}$ = unperturbed root mean square end-to-end distance.

In a good solvent $\alpha > 1$, in poor solvent $\alpha < 1$ and in θ solvent α = unity.

In a very dilute solution of a polymer, the polymer coils stay away from each other and exist independent of each other and exert no effect on each other. In the moderately concentrated solution of the polymer, the coils may entangle with each other. But in the more concentrated solutions, the polymer coils aggregate and remain in swollen viscoelastic gel form.

The determination of root mean square end-to-end distance in theta solvent is important because it is same as the polymer coils in the undissolved state. The value of α gives an idea about the stiffness of the chain.

6.7 Light Scattering Method

This is the most widely used method for the determination of weight average molar mass $\overline{M}_w$. When a beam of light strikes a molecule, the electrons start oscillating at the same frequency as that of the incident light. This causes the development of a transient dipole in the molecules that start acting as the secondary scattering canters by re-emitting in all directions the absorbed energy. This phenomenon is known as scattering.

When a beam of light passes through a solution of polymer, three phenomena can take place—either it will pass through unaltered (however it is less probable), or it will be absorbed (when the energy of radiation matches with the difference in electronic and/or vibrational energies of the polymer molecule—a field of spectroscopy) or it will scatter. Whenever there is a decrease in energy of the incident beam, it emits the radiation simultaneously in all the directions. The scattering by the molecules happens because of the development of an induced instantaneous dipole moment. The total energy of the incident beam is equal to the energies of the transmitted and the scattered beam. If the scattered light has the same frequency as that of the incident beam, it is called the elastic light scattering. If the frequency is different, then it is called inelastic or Raman scattering. The time-averaged intensity of elastically scattered light is termed as static light scattering. It calculates the weight average molecular weight $\overline{M}_w$, the second Virial coefficient, and the radius of gyration of polymers. Dissolved polymer molecules act as discontinuities of sufficient size for light scattering. A dilute solution of the polymer (less than 0.01 g/ml) scatter light from a strong beam of light. The scattered light can be measured by a photometer of suitable design. This is an absolute method. It is used even to determine the weight average molecular weight of polydispersed polymers that have a molecular weight above 10^4. The precision of the results increases with the increase in the molecular weight of the polymer. The intensity of scattered light is related to the concentration and molecular weight of the polymer. The difference in the refractive index of polymer and solvent also affects the intensity of scattered light. The choice of the solvent is made in a manner that the refractive index difference is > 0.1. When a solute is dissolved in different quantities in a solvent, several solutions of different compositions are obtained. These solutions of different compositions have different densities and hence different refractive indexes. Solute particles scatter the light rays. The amplitude of scattering is found to be proportional to the mass M of the particle. It is an experimental fact that the square of amplitude is proportional to the intensity of scattered light. Debye gave an equation in which he related the molecular weight of the solute particle with the intensity of the scattered light that is known as Debye equation. This equation is applicable to the solute particles whose size is smaller than the wavelength of incident light. The equation is

$$\frac{KC}{R_{90}} = \frac{HC}{\tau} = \frac{1}{M} + 2BC \qquad ...(6.41)$$

Where B = second virial coefficient

C = concentration of the solution

R_{90} = Rayleigh ratio at 90° observation angle. R_θ = Rayleigh ratio at any observation angle. So that at 90° observation angle $R_{90} = R_\theta$.

$$R_\theta = \frac{i_\theta r^2}{I_\theta V} \qquad ...(6.42)$$

where I = intensity of the scattered light per unit volume V of scattering material which is observed from a distance r and at an angle θ with reference to the incident beam,

I_θ = intensity of the incident beam

τ = turbidity of the medium

τ is related to the intensity of the incident light. When incident beam passes through the medium of thickness X it intensity reduces from I_0 to I then the value of τ can be given by the following equation $I/I_0 = e^{-\tau x}$).

K and H are light-scattering calibration constants that can be defined by the following equations

$$K = \frac{2\pi^2 n^2 (dn / dC)^2}{\lambda^2 N_A} \quad \text{...(6.43)}$$

n = refractive index of the solution

(dn/dC) = specific refractive index increment (the change of refractive index with concentration)

λ = wavelength of the incident light

N_A = Avogadro's number

$$H = \frac{32\pi^2 n^2 (dn / dC)^2}{3\lambda 4 N_A} \quad \text{...(6.44)}$$

For the calculation for the polymer molecules whose size is larger than the wavelength of incident light the equation following modifications is done.

In the polymer solutions, the polymer molecular coils lie in the form of swollen molecules. The incident beam of light is scattered simultaneously from all the parts of the coil which extend they from A to B. The incident light rays are in different phases when touch the molecular coil. The scattered light rays from different parts of the coil interfere with each other as a result, the intensity of scattered light is found less when observed from an angle θ. It can be said that the θ value varies with θ. Under such conditions equation (6.41) has been modified to give the following equation:

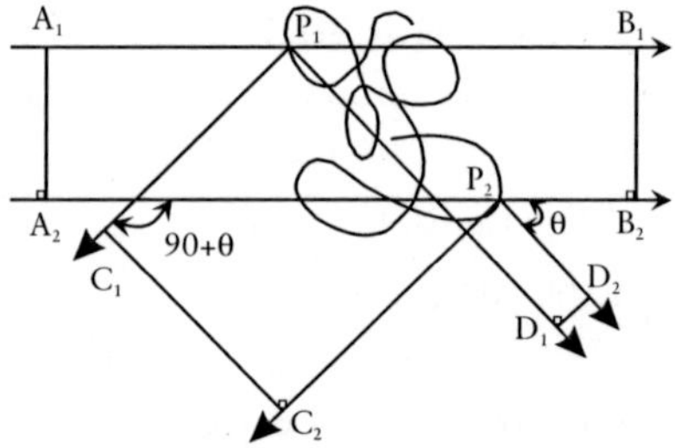

Fig. 6.1: Schematic illustration of the origin of the interference effects which occur between light scattered by two different segments, S_1 and S_2, in the same molecule. The path-length difference ($A_2P_2C_2$ – $A_1P_1B_1$) in the backward direction is greater than the path length difference ($A_1P_1D_1$ – $A_2P_2D_2$) in the forward direction. The path length difference is zero at θ = 0 ($A_1P_1B_1 - A_2P_2B_2$)

$$\frac{KC}{R_\theta} = \frac{HC}{\tau} = \frac{1}{\overline{M}_w P(\theta)} + 2BC \qquad ...(6.45)$$

where $P(\theta)$ = particle scattering factor describing the angular dependence of the scattered light. Its value can be given as follows

$$\frac{1}{P(\theta)} = 1 + \frac{16\pi^2}{3\lambda^2}\left(\overline{S}^2\right)\sin^2\frac{\theta}{2} \qquad ...(6.46)$$

where $(\overline{S}^2)$ = mean square radius of gyration of the polymer coil.

Substituting the value of $P(\theta)$ from equation (6.46) to equation (6.45) we get

$$\frac{KC}{R_\theta} = \frac{HC}{\tau} = \frac{1}{\overline{M}_w} + \frac{16\pi^2}{3\lambda^2}\left\langle\overline{S}^2\right\rangle\sin^2\frac{\theta}{2} + 2BC \qquad ...(6.47)$$

$P(\theta)$ can be considered as a correction factor to the scattered intensity at different values of θ such that $P(\theta)$ = 1 at θ = 1. Similarly putting BC = 0 and C = 0. The equation (6.47) takes the form

$$\left(\frac{KC}{R_\theta}\right)_{C\to 0} = \frac{1}{\overline{M}_w} + \frac{16\pi^2}{3\lambda^2\overline{M}_w}\left\langle\overline{S}^2\right\rangle\sin^2\frac{\theta}{2} \qquad ...(6.48)$$

$$\left(\frac{KC}{R_\theta}\right)_{\theta\to 0} = \frac{1}{\overline{M}_w} + 2BC \qquad ...(6.49)$$

$$\left(\frac{KC}{R_\theta}\right)_{C,\theta\to 0} = \frac{1}{\overline{M}_w} \qquad ...(6.50)$$

On the basis of above equations, we can calculate the molecular weight of the polymer. The calculations can be done either by Debye method or by Zimm method. In the Debye method, the major drawback is one has to find the exact shape of the molecule in solution in order to find $P(\theta)$, which is a difficult task. While, in Zimm method the calculation can be done without knowing the exact shape of the molecule in the solution by adopting a double extrapolation of the plots of (C/R_θ) versus $\sin^2\frac{\theta}{2} + kC$ (where k is an arbitrary constant) to both zero angle of scattering ($\theta = 0$) and at 0 concentration ($C = 0$). When, $P(\theta) = 1$ and BC = 0. The ordinate intercept gives the value of $1/\overline{M}_{wK}$. From the equations (6.48), (6.49) and (6.50), the molecular weight virial coefficient and mean square radius of gyration can be calculated respectively.

With the help of light scattering method, the molecular weights of the polymers in the range 10_3-10_6 can be calculated.

Part – D
Structure and Properties

7 Structure and Properties

7.1 Structure

The structure of the low molecular weight compounds is assumed to be spherical. Collision diameter is a convenient way to have an idea of the molecular size. But owing to the large size of polymer molecules, this was not an acceptable method to express the molecule of a polymer and many schools had to work hard to have a clear picture of polymer structure. Since the polymer chains hardly remain in fully extended form so the flexible polymers have been approximated to be spherical since they become coiled. But the molecular diameter of coiled polymers is very high as compared to the collision diameter of low molecular weight compounds. Our present-day knowledge enables us to express the chain dimensions in different ways.

In order to understand the structure, the simplest example would be the polymer chain made up of carbon-carbon single bond for which we will take the example of polyethylene and a vinyl polymer.

The earliest picture of polymers is the fully extended linear polymer chain. This chain can be represented for polyethylene as follows:

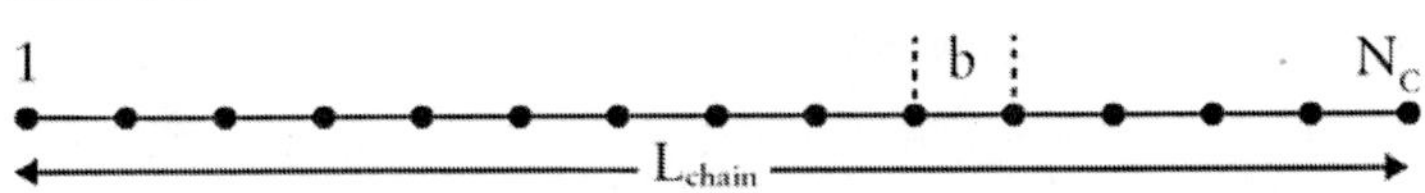

Fig. 7.1

If in this linear chain, there are N_C carbon atoms, then there are N_C-1 number of bonds. All the bonds are assumed to have equal bond length, which is equal to 'b'. This fully extended chain will have maximum length N_b, which was earlier known as the contour length but later on replaced by the term L_{chain}. Then it was realized that the chain cannot be so long because the four valences of carbon are at an angle of 109°28′. So the structure of polyethylene is written as follows:

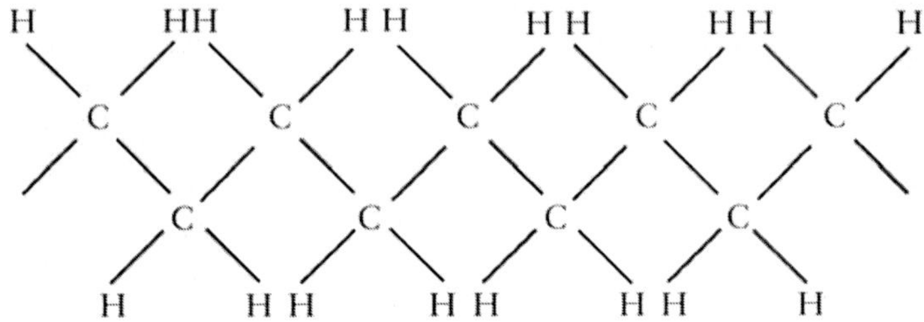

Fig. 7.2

The fully extended chain can be depicted correctly as follows:

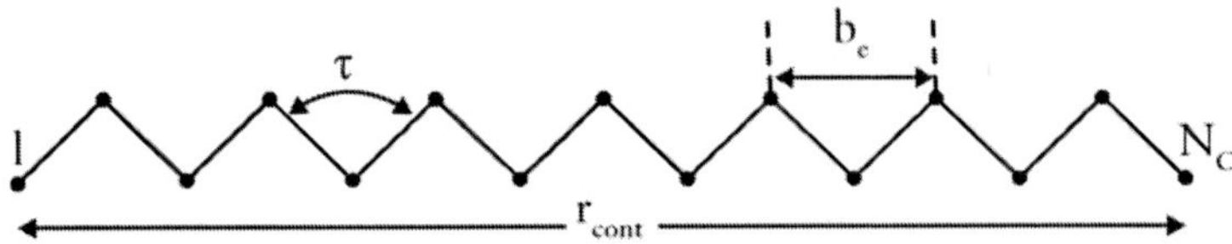

Fig. 7.3

Now, the contour length or the chain length will be written as

$$r_{cont} = Nb \sin (\tau/2) = N_e B_e \qquad ...(7.1)$$

where N_e and B_e are an effective number of bonds and effective bond length and τ is the bond angle. Thus, the contour length is the effective bond length from one end of chain to the other through the centre of the carbon chain. This structural aspect is further improved by considering that the movement of one CRU/ one unit is independent of its neighbours. This improved model of polymer structure is also known as the freely jointed chain.

Statistical Chain Elements

The conformation of a polymer molecule can be viewed as a three-dimensional random walk model. One can start from one end of the polymer chain and put the next step which is equivalent to next monomer unit in a statistically random manner. A mathematical description of each conformation of various polymer molecules corresponds to a particular trajectory of a different particle. It can be explained in terms of diffusion.

theory by comparing the length of a statistical chain element l with the mean free path $\bar{\lambda}$ of a particle moving randomly in space. The number N_{stat} of statistical chain elements, correspond to the j number of collisions. Hence, the length of the polymer chain $L = N_{stat}.l$ corresponds to a diffusion path $s = j\,.\,\bar{\lambda}$ after j collisions. Thus, the diffusion theory can be applied to the polymer structure.

If $P(r)dr$ = probability of finding a diffusing particle after j collisions at a distance between r and $r+dr$ from their original position then,

$$P(r)dr = \text{const } \rho.4\pi r^2 \exp\left(\frac{3r^2}{2j\lambda^2}\right)dr \qquad \text{...(7.2)}$$

and $$\bar{r}^2 = j\lambda^2 \qquad \text{...(7.3)}$$

The equation (7.3) gives the mean square of the displacement of the particle.

When the same is applied to the polymer chain, the probability $P(h)dh$ of finding the distance between the chain ends (end-to-end distance) between h and $h+dh$ becomes

$$P(h)dh = \text{const } 4\pi h^2 \exp\left(-\frac{3h^2}{2N_{stat}l^2}\right)dh \qquad \text{...(7.4)}$$

and $\bar{h}^2 = N_{stat}l^2$ ...(7.5)

The equation (7.5) gives the mean square of the end-to-end distances of the polymer chain.

Chain Dimensions

To determine the size and shape of a polymer molecule has been an interesting and important aspect of research for a polymer scientist. Chain dimension is an important subject of polymer science which deals with the relation of size and shapes of polymer molecules with their chemical structure and the environment around the molecule. The shape of the molecule is directly related to the stiffness of the chain. A flexible chain is more likely to make a coiled structure, but as the rigidity of chain increases it becomes worm-like and finally it turns into a rod-like structure.

The study of a dilute solution of polymers provides a lot of information about their structure.

The possibility of an internal rotation around a single bond gives rise to a countless number of conformations because a chain can be coiled in a variety of ways. The end-to-end distance '*r*' is different for each conformation formed. This distance '*r*' can be estimated from random-walk model of a freely jointed chain.

Average Chain Dimensions

In a dilute solution, with the random thermal motion and flexibility of the chain, the coiled polymer chain is assumed to change its shape continuously. Hence, every chain occupies different volume, where the length of each chain may be different. So, an average dimension needs to be calculated. The two averages are defined depending on the structure of the polymer.

(1) Average root-mean-square is the distance between the chain ends $(\bar{r}^2)^{1/2}$.

(2) Average root-mean-square radius of gyration $(\bar{S}^2)^{1/2}$ is a measure of the average distance between of the chain ends from the centre of gravity of the coil.

Chains are polydisperse and the average over time is to be taken, which is shown by the angular brackets. The bar is put to show the averaging of the various sizes of the chains.

In the case of simple chains, if the excluded volume effects are ignored, then the average root-mean-square distance is related to average root-mean-square radius of gyration by the following equation:

$$(\bar{r}^2)^{1/2} = (\bar{S}^2)^{1/2} \quad ...(7.6)$$

The actual dimensions may vary with the conditions.

Freely Jointed Chain Model

If '*n*' is a number of chain elements joined together with a bond length '*l*', assuming the freedom of rotation about the bond, the chain takes a shape like Fig. 7.4 (a) and (b) where random flight statistics approach is used. The distance between the two ends is r_f of which the value can be calculated from cosine law

$$OB^2 = OA^2 + AB^2 - 2(OA)(AB)\cos\theta \quad ...(7.7)$$

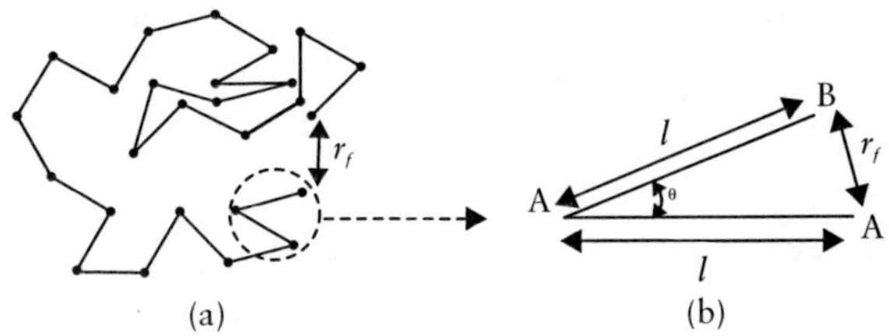

Fig. 7.4 (a) Random walk chain of 26 steps, length l and (b) Cosine law for two bonds.

$$r_f^2 = 2l^2 - 2l^2 \cos\theta \qquad ...(7.8)$$

When the number of bonds 'n' is large and angle θ varies over all possible values, then the sum of all these terms becomes zero. The equation (7.8) will reduce to

$$r_f^2 = nl^2 \qquad ...(7.9)$$

The distance between the chain ends according to this model is proportional to the square root of the number of bonds, which is much smaller than the fully extended chain.

This model was first applied to the diffusion of gas particles and the same was thought for the polymer chain. But in the case of gas every moving particle shifts from its previous position, whereas the polymer segment doesn't follow the Brownian motion like the particle of gas, on the contrary, the segments are placed at fixed positions. This makes the model unrealistic for polymers. The polymer chain occupies a definite volume in space in contrast to the random walk model. The dimensions of polymer chain vary with the bond angles and by the interactions between the chain elements.

There are two types of interactions: 1. Short-range forces of steric repulsion, which occur between neighbouring atoms or groups and are usually forces of steric repulsion caused by the over-lapping of electron clouds; and 2. Long-range interactions comprise attractive and repulsive forces between segments and solvent molecules and between segments; wherein the latter is widely separated in a chain but occasionally approach one another during molecular flexing. These interactions are often termed as the excluded volume effects. In any real polymer chain, two monomers cannot occupy the same space. Even a part of a monomer cannot overlap with a part of the other monomer. This effect is called an excluded volume and plays a far more important role in polymer solutions than it does in solutions of small molecules.

Short-range Effects

The expansion of a polymer chain with covalently bonded is restricted by the valence angles between each chain atom. In general, this angle is θ for a homo-atomic chain, and equation (7.8) can be modified to calculate these short-range interactions.

$$\left\langle \bar{r}^{2} \right\rangle_{of} = nl^2 \,(1 - \cos\theta) / (1 + \cos\theta) \qquad ...(7.10)$$

In equation (7.10), the subscript 'of' stands for the average end-to-end distance of the chains, where the bonds rotate freely around a fixed bond angle. For the simplest case of an all-carbon backbone chain such as polyethylene, θ = 190° and cos θ = –1/3 so that equation (7.10) reduces to

$$\left\langle \bar{r}^{2} \right\rangle_{of} = nl^2 \qquad ...(7.11)$$

This indicates that the polyethylene chain is twice as extended as the freely jointed chain model when the short-range interactions are considered.

The bond rotation is restricted by two factors, viz. chain stiffness and steric repulsion, which is represented by subscript zero, '0'. This means that equation (7.10) must be modified further and becomes

$$\left\langle \bar{r}^{2} \right\rangle_{0} = nl^2 \,\frac{(1-\cos\theta)}{(1+\cos\theta)} \cdot \frac{(1-\langle\cos\phi\rangle)}{(1+\langle\cos\phi\rangle)} \qquad ...(7.12)$$

where ⟨cos ϕ⟩ is the average cosine of the angle of rotation of the bonds in the backbone chain. The parameter $\left\langle \bar{r}^{2} \right\rangle_{0}$ is the average mean square of the unperturbed dimension, which is a characteristic parameter for a given polymer chain.

The structure of real polymer molecules differs from the structure obtained by freely jointed chain model in several ways. Each factor increases the physical dimension of the chain and consequently the end-to-end distance. The restriction imposed on bond angle and free rotation about chain bonds increases the chain dimension. The dimension that results by considering the above two factors using random flight model is known as the unperturbed dimension of the polymer chain and it is represented by $\left\langle \bar{r}_0^2 \right\rangle^{1/2}$.

Long-range Effects

The corrections made by short-range interactions are still insufficient to explain the existence of conformations in which two widely separated chain segments occupy the same space. This is because in real chains each segment occupied a volume from which all other segments are excluded. The concept of free volume appeared which forced to develop the 'self-avoiding walk' model. The effect of the long-range interactions resulted in further expansion of unperturbed dimension of chain. This expansion is represented by an expansion factor α. The value of α depends on the nature of the solvent. From here the concept of thermodynamically 'good' solvent for a particular solvent arose.

The dimension of a polymer molecule and that of the freely jointed chain model are related by the following equation:

$$\left\langle \bar{r}_0^2 \right\rangle^{1/2} = N \left\langle \bar{r}_f^2 \right\rangle^{1/2} \qquad \text{...(7.13)}$$

$\left\langle \bar{r}_0^2 \right\rangle^{1/2}$ = actual root mean square end-to-end distance of polymer molecule known as an unperturbed dimension of the molecule.

N = a number which is > 1

$\left\langle \bar{r}_f^2 \right\rangle^{1/2}$ = root mean square end-to-end distance of polymer molecule by random flight model.

7.2 Glass Transition Temperature (T_g) and Melting Point (T_m)

Introduction

In contrast to low molecular weight compounds, which exist as solid, liquid and gas, the polymers exist only as solid and liquid. One of the characteristic features of polymers is that they decompose before their melting point. All the polymers become brittle and transparent glassy solids at low temperatures, for example, an article made up of ordinary rubber at –70°C becomes like glass and if dropped on the floor, would break like glass. There are polymers which degrade on heating, and there are others that do not and rather become viscous liquids when heated. The polymers with ordered regions are termed as crystalline polymers. Polymers that have a random arrangement of chains are called amorphous. But, in real, polymers are seldom cent percent crystalline. The extent of crystallization is represented in terms of percentage crytallinity. The chains in some polymers are ordered in three dimensions with an alternate disordered arrangement. Such polymers are called crystallites and have semi-crystalline structures, e.g. spherulites. If a solid polymer, whether amorphous or partially crystalline is heated, it increases in volume.

We have already seen that one can synthesize different samples of a polymer that have different molecular weights. It is interesting to note that a typical amorphous polymer such as polystyrene, can exist in several states depending on its molecular weight and temperature. At room temperature, some of the amorphous polymers are hard and rigid glassy solids while some are soft, flexible and rubber-like. At low temperature, the amorphous polymer is a hard glassy material and on heating, it melts into a viscous liquid through a rubbery state. The temperature at which a hard glass polymer becomes a rubbery material is called the glass transition Temperature which is represented by T_g. This soft state of the polymer is a rubbery or viscoelastic state. If a graph is plotted between temperature and volume or enthalpy, then separate regions of dramatically different polymer properties are obtained below T_g. When an amorphous polymer is heated above, it displays a rubbery region which is soft and pliable. But a sample of the same polymer with high molecular weight becomes a highly viscous liquid on further heating and starts flowing, which is another viscoelastic state. This temperature is known as the Flow temperature represented and is by T_g.

The thermodynamic transitions are of two types: the first order and the second order. The first order transitions are characterized by an abrupt change in a fundamental thermodynamic property, e.g. enthalpy H, volume V, etc. While in the second order transitions the first derivative of the properties change, e.g. the heat capacity C_p, volume thermal expansion coefficient α

$$C_p = \left(\frac{\partial H}{\partial T}\right)_p \text{ and } \alpha = \frac{1}{V}\left(\frac{\partial V}{\partial T}\right)_p$$

During the first order transition of polymers at T_g, enthalpy and volume change abruptly, while in the second order transitions C_p and change.

Free Volume Concept

The free volume concept in the case of polymers is the space in the liquid and solid samples of polymers which is not occupied by the polymer molecules. This remains as an empty space more between the liquids than the solids. This is the reason that the molecules change their conformations freely in the liquid state. When the temperature is reduced, the thermal energy responsible for molecular motion also reduces. A stage is reached when the sample does not have enough free volume to allow molecular rotation or translation. The temperature at which this happens is T_g.

A crystalline polymer differs from an amorphous one. The crystalline polymer turns rubbery through a flexible crystalline state. On further heating, this rubbery material is converted into a viscous liquid. The temperature at which it is converted to the liquid state is known as the melting point T_m. The volume of a crystalline solid changes at the crystalline melting point T_g. This behaviour is contrasted with that of an amorphous polymer in which the flow occurs at flow temperature T_f. Amorphous polymers do not have melting point. The temperature at the point of deflection in the slope, varies with the cooling rate. The rubbery polymers convert to the liquid state on heating, but this phenomena is not observed in the low molecular weight compounds such as H_2O, C_2H_5OH, etc. The low molecular weight compounds have a sharp melting point. As the molecular weight of the polymer increases, the value of both T_g and T_f also increases.

Polymers have a certain range of temperature in which they are liquid. Polymers below the lower temperature limit are brittle, and above the upper-temperature limit, they first become soft, then degrade and finally decompose with the increase in the temperature. At very low temperatures, all the polymers are hard rigid solids, as the temperature increases they become a viscous liquid unless they do not degrade.

Attempts are continuously made to extend the working temperature of the polymers. Introduction of metal atoms increases the thermal stability of the polymer. Inorganic polymers and polymers containing organo-metallic bonds have good heat resistance. The introduction of aromatic and heterocyclic rings improves the thermal stability.

Polymers have two main types of thermal transitions:

(1) Crystalline melting point (T_m)

(2) Glass transition temperature (T_g)

The glass transition temperature is the temperature at which the amorphous part of the polymer becomes glassy. Completely amorphous polymers exhibit only T_g. Completely crystalline polymers have only T_m. But because most polymers are both partly crystalline and amorphous, thus they show T_m as well as T_g.

The schematic representation of thermal response change of specific volume (v) of all the three types of polymers are given as follows:

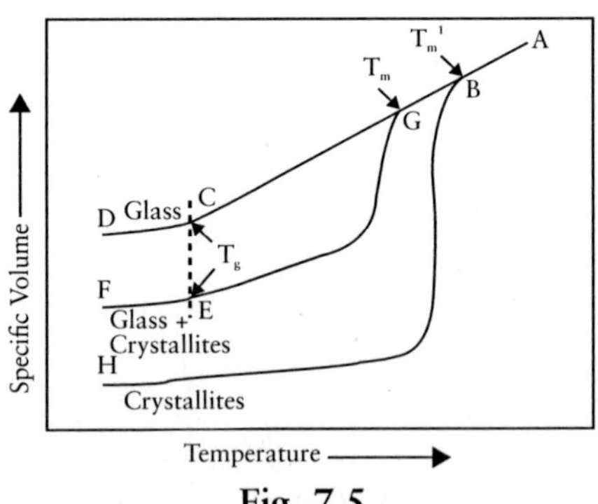

Fig. 7.5

(i) Completely amorphous sample (D-C-G-B-A),

(ii) A semi-crystalline sample (F-E-G- B-A), and

(iii) A perfectly crystalline material (H-B-A).

If an amorphous solid is heated, it will follow the curve D-C-G-B-A. At point B, it becomes a rubbery mass and at G, into a viscous liquid. The solid amorphous polymer increases in volume when heat is given to it. Curve D-A is obtained. In the region D-C, the polymer is glassy. At point C, it starts becoming soft, ductile and becomes rubber-like. This point is known as glass transition temperature (T_g). When heat is continuously supplied it becomes a viscous liquid.

In a crystalline polymer, no T_g is obtained because of the absence of disordered regions. On further heating, it follows the curve H-B-A and at point B melting of the polymer occurs and it becomes a viscous liquid. Point B is the melting temperature of a perfectly crystalline polymer of high molar mass represented by $T_m°$. The semi-crystalline polymer follows the curve F-E-G-B-A. It shows a melting range from G-B, i.e. $T_m - T_m°$. The imperfection of semi-crystalline structure of polymers decreases the melting point of the polymer from $T_m°$ to T_m. The semi-crystalline polymers have T_g at point E. In all the cases from B-A, all the polymers are viscous liquids.

It is interesting to note that at the glass transition temperature T_g, the ratio of the volume of the polymer chain segment to the free volume associated with that segment in which it moves, is the same for all the polymers. The value of T_g varies with the variation of chemical structure of the polymer. If there are structural changes in the polymer molecule, the segmental volume would also change. This also makes a change in the free volume per segment, as a consequence of which T_g also changes. T_g increases with increasing chain stiffness (which is a result of adding hydrogen bonding, the addition of stiff or bulky side groups, the introduction of steric hindrance, etc.) to get the same free volume level. Similarly, T_g decreases with the addition of low molecular weight additives, as they increase free volume.

One of the most convenient methods to measure T_g is to use a Differential Scanning Calorimeter (DSC).

Low molecular weight polymers have a greater number of chain ends compared to those of high molecular weight in a given volume. The less restrained chain ends have a more libral mobility at a given temperature, which results in a lower value depends on molecular weight of a polymer and is correlated by

$$T_g = T_g^{\infty} - \frac{K}{\mu_n} = T_g^{\infty} - \frac{K}{\overline{M}_n} \qquad ...(7.14)$$

where $T_g^{\infty} = T_g$ value of a fictitious sample of the same polymer of infinite molecular weight,

μ_n = number of average chain length of the polymer sample,

K = a positive constant that depends on the nature of the material.

In the case of copolymers, the glass transition temperature lies between the T_g values of both the homopolymers.

In homogeneous copolymers, the value of T_g is intermediate between the values of both the monomers. If the glass transition temperatures are T_{g1} and T_{g2} for the individual homopolymers are known it is possible to calculate the T_g of copolymer by using the following equation

$$T_g = V_1 T_{g1} + V_2 T_{g2} \qquad ...(7.15)$$

where V_1 and V_2 are the volume fraction of components 1 and 2, respectively (Fig. 7.6, line 1).

In random or alternating copolymers, e.g. formed by methyl methacrylate – acronitrile T_g can be calculated by the following equation:

$$\frac{1}{T_g} = \frac{w_1}{T_{g1}} + \frac{(1 - \mathrm{w}_1)}{T_{g2}} \qquad ...(7.16)$$

where w_1 = weight fraction of one of the monomers present in the copolymer,

T_{g1} = glass transition temperature of monomer 1 and

T_{g2} = glass transition temperature of monomer 2.

The glass transition temperature varies with the composition of the copolymer.

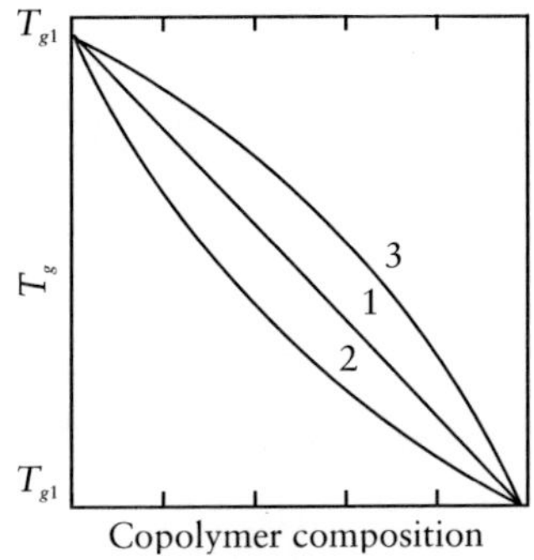

Fig. 7.6: Schematic diagram showing variation in glass transition temperature with copolymer composition

Because of interaction between the chains, the value of T_g increases (Fig. 7.6, line 3). If the disorder increases because of random distribution the free volume increases consequently the value for T_g decreases (Fig. 7.6, line 2).

The polymer molecule can have four types of motion:

(1) Translational motion of the entire molecule.

(2) Long cooperative wriggling motion of 40 to 50 C—C bonds of the molecule which causes flexing as well as uncoiling of the polymer chain.

(3) Short cooperative motion of 5 to 6 C—C bonds of the molecule.

(4) Vibration of carbon atoms in the polymer molecule.

Below T_g, long and short cooperative wriggling and translational motions are quenched. The polymers whose T_g value is less than room temperature are rubbery. In the rubbery state, only the translational motion is quenched.

8 Polymer Solutions

Thermodynamic properties of a polymer solution depend on how 'good' the solvent is for the polymer, as well as, on the polymer itself. The interaction between the solvent and the polymer and degree of polymerization dictate these properties.

Introduction

There is a direct relationship between the synthesis of a polymer and its molecular properties. Hence, a detailed and careful method of synthesis provides a polymer with the required characteristics needed by the end user. This makes polymers more and more useful in our daily life and enhances the lifestyle fulfilling specific daily needs with its unique properties.

The study of relationship between molecular properties and bulk properties of polymers is the very foundation of polymer science. In order to establish this relationship, one has to characterize the polymer. The molecular characterization of a polymer includes determination of basic structure, average molecular weight, molecular weight distribution, etc. Most methods developed for characterization use a dilute polymer solution (< than 20 g dm^{-2}). So, one has to understand various aspects of the polymer solution.

8.1 Thermodynamics of a Solution

Ideal mixture of small molecules (solutions of simple molecules)

Ideal solutions are a mixture of two types of molecules: the solvent '1' and the solute '2'; which are comparable in size. In this solution, the force of attraction of 1-1 molecules and

2-2 molecules should be equal to the force of attraction of 1-2 molecules. Under this condition the $\Delta H_m = 0$ and implies that the rotational, vibrational and electronic entropies do not change. The ΔS_{mix} depends only upon the Configurational (combinatorial) entropy change ΔS_{mix}^{combi}. When two different substances are mixed, the number of distinguishable spatial structures of equal energy increase, hence ΔS_{mix} is always positive. As a result, ΔG_{mix} is negative and the solution forms. Statistical mechanics can easily relate the number of these degenerate spatial structures with entropy by applying Boltzmann equation.

a b

Fig. 8.1: Schematic representation of a liquid lattice. (a) mixture of molecules of equal size, (b) mixture of solvent molecules with a polymer molecule showing the connectivity of polymer segments

$$S = K\ln \Omega \qquad ...(8.1)$$

where k = Boltzmann's constant and

Ω = total number of distinguishable degenerate arrangements.

Equation (8.1) can be applied to the formation of an ideal solution.

$$\Delta S_{mix}^{combi} = k[\ln\Omega_{12} - \ln\Omega_1 + \ln\Omega_2] \qquad ...(8.2)$$

where Ω_1 = the total number of distinguishable spatial arrangement of molecules in a pure solvent, Ω_1 = in solute and Ω_{12} = in ideal mixture.

If there are N identical molecules to be arranged in a lattice containing N cells, then there will only be one possible arrangement and hence $\Omega_1 = \Omega_2$. Consequently, the equation (8.2) becomes

$$\Delta S_{mix}^{combi} = k\ln \Omega_{12} \qquad ...(8.3)$$

By using Stirling equation for ideal mixing of N_1 molecules of a solvent and N_2 molecules of a solute in a lattice with

$(N_1 + N_2)$ cells, the entropy can be calculated. In terms of number of moles m and mole fraction X, the following equation for entropy is obtained, where $R = kN_A$

$$\Delta S_{mix}^{combi} = -R[n_1 \ln X_1 + n2 \ln X_2] \qquad ...(8.4)$$

where $n_1 = \dfrac{N_1}{N_A}$, $n_2 = \dfrac{N_1}{N_A}$, $X_1 = \dfrac{n_1}{n_1 + n_2}$, $X_2 = \dfrac{n_2}{n_1 + n_2}$

It has been assumed that

(1) During mixing, change in volume is zero, i.e. $\Delta V_{mix} = 0$
(2) All the molecules are of the same size
(3) All possible arrangements have the same energy, i.e. $H_{mix} = 0$
(4) The motion of the components about their equilibrium positions remains unchanged on mixing.

The free energy of this solution is:

$$\Delta G_{mix} = RT[n_1 \ln X_1 + n_2 \ln X_2] = -T\Delta S_{mix} \qquad ...(8.5)$$

This equation shows that mixing in ideal systems is spontaneous and is solely driven by entropy change. Since most of the solutions are not ideal, so modification is to be done in simple lattice theory. In the case of small molecules, the value of $\Delta H_{mix} = 0$, that is non-athermal mixing but ΔS_{mix} is ideal and the solution becomes non-ideal. There is a third type of situation in which ΔS_{mix} and ΔH_{mix} deviate from the ideal values. This is called an irregular solution. The polymer solutions lie in this category.

The polymer solution comes under the category of irregular solutions because of the finite heat of mixing and the large difference between the size of the molecules of the solute and the solvent. It is the long chain of polymers which is the cause of the deviation from the ideal entropy of mixing and consequently from Raoult's law. Polymer solutions obey Raoult's law only to some extent under the condition of extreme dilution. Because of the long chain character of the polymers, the method of calculation of the number of ways of arranging polymer chains and solvent molecules in the lattices becomes different from the arrangement of small solute molecules and solvent molecules in the lattices.

Flory and Huggins independently proposed a modified theory which has taken into consideration the great difference in the size of the solvent molecules and the polymer (solute) molecules. They have also considered the inter-molecular interactions.

First of all from the point of view of thermodynamics, the situation in solutions of simple molecules can be solved as follows:

A solution of two components 1 and 2 is formed by mixing n_1 and n_2 moles (N_1 and N_2 molecules) at constant temperature T and pressure P. The total free energy of this system is:

$$G_{12} = n_1 {}^{\circ} + n_2 G_2^{\circ} + n_1 RT \ln \left(p_1 / p_1^{\circ} \right) + n_2 RT \ln \left(p_1 / p_2^{\circ} \right) \quad ...(8.6)$$

where G_1° and G_2° = free energies of pure components 1 and 2,

p_1° and p_2° = vapour pressures of pure components 1 and 2,

p_1 and p_2 = partial vapour pressures of the pure components 1 and 2 in the solution respectively.

According to Raoult's law

$p_1 = x_1 p_1^{\circ}$ and

$p_2 = x_2 p_2^{\circ}$

where x_1 and x_2 are the mole fraction of components 1 and 2.

We know $x_1 + x_2 = 1$ or $x_1 = 1 - x_2 = n_1 / n_1 + n_2$

So, the equation (8.6) becomes

$$G_{12} = n_1 G_1^{\circ} + n_2 G_2^{\circ} + n_1 RT \ln x_1 + n_2 RT \ln x_2 \quad ...(8.7)$$

$$\text{or} \quad G_{12} - \left(n_1 G_1^{\circ} + n_2 G_2^{\circ} \right) = n_1 RT \ln x_1 + n_2 RT \ln x_2$$

$$\text{or} \quad G_{\text{mix}} = n_1 RT \ln x_1 + n_2 RT \ln x_2$$

$$\text{or} \quad G_{\text{mix}} = RT \sum n_i \ln x_1 \quad ...(8.8)$$

Differentiation of equation (8.8) with respect to temperature T gives

$$\frac{\delta \Delta G_{\text{mix}}}{\delta T} = n_1 RT \ln x_1 + n_2 RT \ln x_2 \quad ...(8.9)$$

$$-\frac{\delta \Delta G_{\text{mix}}}{\Delta T} = \Delta \sin x = - n_1 RT \ln x_1 - n_2 RT \ln x_2 \quad ...(8.10)$$

Combining equations (8.4) and (8.9), we get

$$\Delta G_{\text{mix}}^{\text{ideal}} = \Delta S_{\text{mix}}^{\text{ideal}}$$

or

$$\Delta H_{\text{mix}}^{\text{ideal}} = 0 \qquad ...(8.11)$$

However, for non-ideal or real solutions, the activity coefficient terms comes into picture and we have

$$p_1 = x_1 y_1 p_1^{\circ} \qquad ...(8.12)$$

$$p_2 = x_2 y_2 p_2^{\circ} \qquad ...(8.13)$$

where y_1 and y_2 are the activity coefficients for components 1 and 2

By introduction of equations (8.12) and (8.13), we get

$$\Delta G_{\text{mix}} = RT \sum n_i \ln x_i + RT \sum n_i \ln y_i \qquad ...(8.14)$$

$$\Delta S_{\text{mix}} = -R\sum n_i \ln x_i - R \sum n_i \ln y_i - RT\sum n_i(\delta l y n_i / \delta \mathrm{T}) \qquad ...(8.15)$$

$$\Delta H_{\text{mix}} = RT^2 \sum n_i (\delta \ln y_i / \delta \mathrm{T}) \qquad (8.16)$$

More the value of ΔH_{mix} is negative, higher will be the dissolution. Presence of attractive forces like hydrogen bonding, etc. between polymer and solvent makes ΔH_{mix} more negative. This causes liberation of more heat from the system during dissolution. If ΔH_{mix} is positive that means the polymer and the solvent want to remain in each other's company. Thus, $\Delta \mathrm{H}_{\text{mix}}$ is either less than $T\Delta S$ or it is greater than $T\Delta S$. If $\Delta H_{\text{mix}} > T\Delta S$ then dissolution does not take place, but, if $\Delta H_{\text{mix}} < T\Delta S$ the dissolution takes place because the resultant sum of equation ΔG_{mix}, becomes negative. When the value of ΔH_{mix} is small, it is ΔS_{mix} which determines the sign of ΔG_{mix}.

8.2 Aggregation in Polymers

First, let's look at the structure of a solid crystal of a low molecular weight substance. Each molecule (or ion) positions itself in the crystal lattice in equilibrium, where it can only vibrate, but has no translational movement or Brownian moment. This definite positioning of molecules is repeated in all the three directions of the crystal and the crystal is known to have a long-range order (lro). This molecule is rigid and inflexible to any external force applied and can transfer the force to a nearby molecule. In order to define this crystalline solid one needs to consider its three aspects:

(1) Position and condition of the molecules constituting the crystal
(2) Geometry of crystal lattice
(3) Response to the external force applied.

When a solid is heated, its kinetic energy increases and its molecules start vibrating at their fixed positions. If there is a continuous supply of heat, the molecules vibrate more vigorously along their fixed equilibrium positions and reach a stage when they start moving from their fixed position. This is known as the Brownian movement. If the supply of heat is further continued, then, this movement intensifies and as a consequence, the long-range order of the crystal would be lost. At this stage, if an external force is applied, it will not be transferred to the neighbouring molecule and the substance starts flowing. This is called the liquid state. There are two states in low molecular weight compounds: with Brownian movement and without Brownian movement. The state without Brownian movement has long-range order and stress could be transferred. The state with the Brownian movement has short-range order and is able to dissipate energy but stress cannot be transferred. The state with Brownian movement is known as the liquid state. The temperature at which the solid changes to liquid state is known as the melting point T_m.

Now let's look at the case of polymers. If we take a simple case of polyethylene, which is a solid at room temperature and behaves like a low molecular weight compound in many of its properties. But as the temperature increases, it starts behaving differently because it is made of chain segments. These chain segments start moving in one part of the solid mass while in other parts they remain unmoved. So the motion of chain segments is localized. At the same time some of the chain segments in a single chain move while the others do not. There are two clearly distinguishable movements inside the solid mass.

(1) The localized mobility of the segments, i.e. segmental motion

(2) The mobility of the molecular chain as a whole is known as the molecular motion.

They are both called Brownian movements. The former is Internal or Micro Brownian movement (IBM) and the latter is External or Macro Brownian movement (EBM). When the heat is supplied to a solid polymer, first IBM gets activated and if heat supply is continued, the EBM also gets activated. As a result, the flow of the polymer starts in the form of a highly viscous liquid. In order to activate IBM alone, one has to make a careful transfer of heat. Once, only IBM is activated, polymer segments would be in liquid state but the polymer molecule as a whole will remain in solid state. This is known as the rubbery state. In this situation under the influence of applied stress, polymer behaves as a viscous fluid as well as an elastic solid and undergoes viscoelastic deformation.

8.3 Configuration of Polymers

The polymers, whether natural or synthetic, can be divided into two groups

1. Rigid polymers
2. Flexible polymers

However, there are a number of polymers whose molecules do not fit in either of the class. In addition to this, there are a number of polymers which shift from group 1 to group 2 with the increase of the temperature.

Proteins have rigid structure and belong to the first group. The presence of intramolecular hydrogen bonding provides inflexibility to a molecule. As a result, these molecules have a definite shape.

The synthetic linear polymers are the example of flexible polymers. The flexibility of the molecule is due to the freedom of rotation of mer-mer bond (the bond between segments). As a result, these molecules coil up to have a more compact configuration in solid, liquid as well as in the solution phase. An important thing about these configurations is their changeability with time with reference to an average configuration.

To understand this first we should understand the chain dimensions. The subject of chain dimension is concerned with the relationship of size and shape of a polymer with the molecular environment. The chemical structure of a polymer molecule affects the chain stiffness which in turn determines the shape of the polymer molecule. Mostly, the polymer molecules have a flexible backbone (simplest case is of carbon backbone) and tend to coil. With the increase in the stiffness of the backbone, the molecule first becomes worm-like and finally it becomes rod-like.

8.4 Process of Polymer Dissolution

The solubility characteristics of low molecular weight substances and polymer in a solvent are entirely different.

	Polymer	Low molecular weight substance
Time taken for dissolution	very high	very low
Viscosity	very high even at low concentration	very less

It is necessary to understand the phenomenon of dissolution in both the cases.

We take water as a solvent and a crystalline salt A^+B^- as solute and stir it to dissolve. Gradually, the crystal becomes smaller until it finally disappears. The phenomenon from the point of view of Chemistry is that the ions leave the crystal lattice and disperse in water as the size of molecules of the solvent and the solute are comparable. The surrounding water molecules are able to break the electrostatic forces between the ions in the crystal. In this case, the solute-solvent interactions are stronger than the solute-solute and solvent-solvent interactions. The two phases, viz. solid and liquid, almost instantaneously become one single phase, i.e. the solution phase. If the solution was not stirred then it would have taken longer.

In contrast to this, if we take a polymer like polyvinylalcohol (since PVA is soluble in water) and put it in water and stir it well, nothing would happen. After a long time, the PVA crystal would start to swell and their size would go on increasing. But, if this mixture is stirred faster, PVA would slowly start dissolving. We see that there are two stages during the dissolution of PVA in water: swelling and dissolution. The giant structure of PVA with a large number of chain segments in the polymer chain is responsible for this. These chain segments lie in the form of tightly folded coils. These coils also interpenetrate each other because various types of intermolecular and intramolecular cohesive and attractive forces play their role in keeping these chains together. These forces are responsible for the difficulty in the solubility of the polymers. But, even if the solubility of a polymer with linear flexible chain is considered, factors like forces of attraction and dispersion would also start playing their role. The extent of solubility depends on the chemical nature of the polymer chain, the polarity of groups attached, the solubility parameter of polymer, etc. The solubility of polymer depends on the increase of the solute-solvent interaction than that of solute-solute and solvent-solvent interaction. If this factor predominates the penetration of solvent molecules into the segment, then there are chances that the segment-segment contacts break causing an interaction of each segment with solvent molecules. This situation is known as the solvation.

The time taken for the slow process of swelling and dissolution is sometimes known as pseudo induction period. The molecules of the solvent enter into the polymer matrix and the volume of the polymer increases, which is known as the swelling of the polymer. Once swelled, the polymer chains start diffusing into a solvent resulting in the formation of the solution. Though the polymer molecules remain expanded in the solution, nevertheless they are still somewhat coiled. This coiling is such that the solvent molecules remain trapped inside the coils. The trapped solvent is known as 'bound solvent' and the rest of the solvent as 'free solvent'. However, the bound solvent is in dynamic equilibrium with the free solvent. The apparent volume occupied by the expanded coil is known as 'hydrodynamic volume' of the polymer.

In the case of solutions of low molecular weight compounds, in the absence of swelling, the intermolecular friction does not increase much, as a result of which the solute particles have high mobility. But in the case of polymers the big chains swell up, hence, the increase in volume is high, as a result of which the intermolecular friction also increases. This intermolecular friction hampers the mobility of polymer molecules which in return increases enormously the viscosity of the solution.

8.5 Polymer Solution

A solution can be defined as a homogenous mixture of two or more substances which are mixed on a molecular scale. The simplest example is when a substance is dissolved, it is called the solute and the other in which dissolution takes place is the solvent. From the thermodynamics point of view, a solution will form at constant pressure P and temperature T, only when the Gibbs free energy G_{12} of the solution is less than the sum of free energies of pure components G_1 and G_2. Gibbs free energy of mixing G_m should be less than zero, i.e.

$$\Delta G_m = (0) = G_{12} - (G_1 + G_2) \qquad ...(8.17)$$

Gibbs free energy is related to enthalpy and entropy by the relation

$$G = H - TS$$

where H = enthalpy, T = temperature on absolute scale and S = entropy. This equation gives a useful equation

$$G_m = \Delta H_m - T\Delta S_m \qquad ...(8.18)$$

where ΔH_m = enthalpy of mixing and ΔS_m = entropy of mixing.

When dealing with polymer solutions the thermodynamic properties are analyzed by using the equation (8.19).

$$\mu_i - \mu_i^0 = RT \ln a_1 \qquad ...(8.19)$$

where μ_i = chemical potential of component 'i'; μ_i^0 = chemical potential of component 'i' in its standard state; and $a_1 = y_1 X_1$, where activity coefficient of solvent and mole fraction of the solvent.

It is only the contribution of the solvent that is considered in this equation. The solvent contributes in two ways:

(1) Ideal contribution
(2) Non-ideal contribution, also known as excess contribution

Hence, ideal contribution is:

$$(\mu_i - \mu_i^0)^{ideal} = RT \ln X_1 \qquad ...(8.20)$$

and non-ideal contribution is:

$$(\mu_i - \mu_i^0)^E = RT \ln Y_1 \qquad ...(8.21)$$

where superscript E is excess or non-ideal contribution.

8.6 Molecular Structure of Flexible Linear Molecule in Dilute Solution

Let us take a chain of carbon atoms in which free rotation about each C—C bond can take place with the restriction that C—C—C bond angle should remain fixed at 109°28′, where the C—C internuclear distance is 1.54 Å. There is no alternate arrangement till there are two carbon nuclei C_1 and C_2. As soon as the third carbon atom enters into the chain, it can be placed anywhere on a circle obtained by rotating C_2—C_3 bond at a constant angle of 109°28′ to the C_1—C_2 bond. When fourth carbon atom is introduced it can be placed on a similar circle drawn relative to C_2—C_3 bond. So there are a large number of positions of the fourth nuclei in relation to the fixed axes in space.

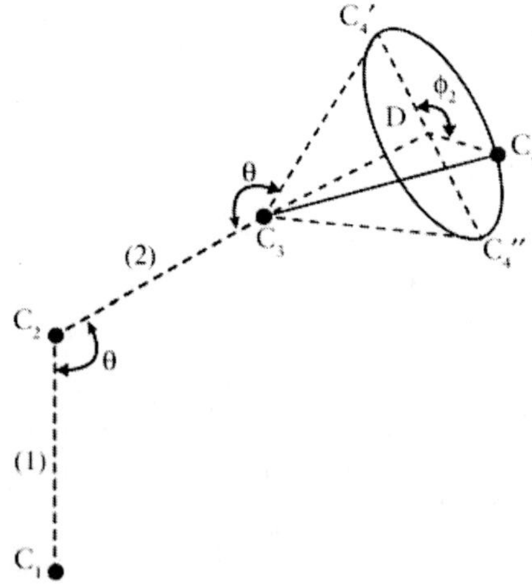

Fig. 8.2

Suppose the carbon atoms C_1, C_2, C_3 and C_4 lie in the plane of the paper, the following four configurations will be obtained.

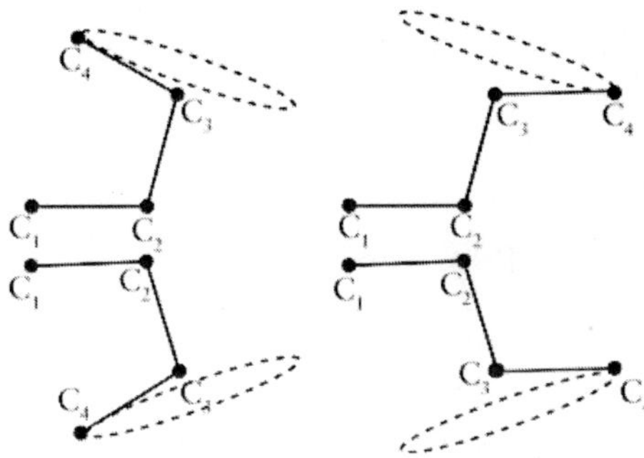

Fig. 8.3

Since these four structures are formed by the rotation of C_2—C_3 bond, these structures are inter-convertible. The number of structures increases enormously as the number of carbon atoms increase. Moreover, the positions are occupied by the carbon nuclei and not merely by bonds of the electrons. Actually, in reality, there is no possibility of occupying the same position by another carbon atom. In addition, the rotation around each C—C bond is hindered by repulsive forces between the substituents on the adjacent carbon atoms (this effect becomes more pronounced if a bulky substituent group is present). Finally, we can see that the polymer molecule has no fixed configuration of the chain because of the continuous rotation around C—C bonds and the chain adopts a succession of different poses. The chain prefers the configurations in which the repulsive forces between groups are low. Accordingly, the dimensions of the molecule can be characterized by taking the average of the distance between the two ends of the chain. A long chain assumes a coiled structure in the presence of secondary valence forces throughout the chain. As, it is difficult to find the length of the polymer chain, thus, measuring the end to end distance is the preferred method. In order to understand the various conformers of polymers, the random flight technique and classical statistics are helpful.

It is possible to describe an isolated polymer chain only when it is in a solution. If the solution is very dilute these polymer chains disengage from each other and the solvent molecules come in between them. Then the internal movement of each chain takes place independently. As a result, one can find some molecules to be tightly coiled, some expanded and numerous others in between stages. In fact, there will not be many molecules having fully extended chains as that will lead to high degree of order in the solution.

Mostly, it is the polymers with flexible chains that remain coiled in the solution. Sometimes even the most rigid polymer chain does not remain in fully extended form due to the solvent's effects. The polymer chains either have full contour length which is the product of the length of monomer (b) and number of monomers (*N*) or else they have extended length which is $Nb\sin\theta/2$, where is the bond angle between the chain links. The bond of random coils has chain segments (which are formed of some monomer units) which rotate at fixed angles, to form various conformations that can be calculated by random flight probability. The chain ends of the various conformations are separated by a distance r_0, which is the end-to-end distance. The end-to-end distance is better expressed as root mean square end-to end distance $(\overline{r_0^2})^{1/2}$.

The scene in a dilute solution is that the separated molecular chains are in various extended positions and which is why each end to end distance varies. We can calculate the average end-to-end distance $\bar{r}$ by measuring the end to end distance of each separate molecule. In order to calculate it, let's suppose that there are n_1 molecules with end to end distance r_1, n_2 molecules with distance r_2 ... n_i molecules with distance r_i and so on, then the average end to end distance r would be

$$\bar{r} = \frac{n_1 r_1 + n_2 r_2 + \ldots + n_i r_i}{n_1 + n_2 + \ldots + n_i} = \frac{\sum n_i r_i}{\sum n_i} \qquad \ldots(8.22)$$

It has been found that instead of average end-to-end distance, it is better to calculate root mean square end-to-end distance $((\overline{r_0^2})^{1/2})$.

$$\text{So } (\overline{r^2})^{\frac{1}{2}} = \left\{\frac{n_1 r_1^2 + n_2 r_2^2 + \ldots + n_i r_i^2}{n_1 + n_2 + \ldots + n_i}\right\}^{1/2} = \left\{\frac{\sum n_i r_i^2}{n_i}\right\}^{1/2} \qquad \ldots(8.23)$$

It should be noted here that the same polymer will have a different root mean square end-to-end distance when dissolved in different solvents.

If the concentration of polymer solution is increased, the polymer chain starts getting entangled. However, chains still may have a high degree of flexibility but the segmental motion of one chain does not remain independent of the other chain that influences the flow properties of these solutions. That is the reason why flow properties of polymer solutions are different from the solutions containing simple solutes. In the case of branched polymers, the molecule becomes like a ball. Thus, its radius of gyration $(\overline{S}_0)$, gives the idea of the structure. Better results are obtained if instead of S_0 the root mean square value $(\overline{S_0^2})^{1/2}$ is used. This is the distance from the centre of gravity of molecular coil to the chain end. The root mean square end-to-end distance $(\overline{r_0^2})^{1/2}$ and radius of gyration root mean square distance $(\overline{S_0^2})^{1/2}$ are related as follows:

$$(\overline{r_0^2})^{1/2} = 6(\overline{S_0^2})^{1/2} \qquad ...(8.24)$$

Several different types of forces develop when a polymer is added to a solvent. One of them is the cohesive force between the chain segments of a polymer molecule which tightens further the polymer coil. Other is a force of attraction between the chain segments and the solvent molecules, which loosens up the polymer coil. It is the resultant of these two forces that the actual size of the polymer coil is determined. If the solvent attraction is more the polymer chain expands and $(\overline{r_0^2})^{1/2}$ become larger, whereas it decreases if the solvent attraction is less. In the case when solvent does not interact with the polymer at all, the polymer coil acquires an unperturbed dimension and is known as Theta solvent or Flory solvent. The temperature at which the polymer chains are on the verge of precipitation is called Theta temperature or Flory temperature.

In practice, the average dimension of polymer chain becomes different from the unperturbed dimension. In good solvents

$(\overline{r_0^2})^{1/2}$ becomes equal to $\alpha(\overline{r_0^2})^{1/2}$, where is the chain expansion factor. The value of α can be calculated by comparing end-to-end distance in any solvent to end-to-end distance in theta solvent. The root mean square end-to-end distance of the un-dissolved coil of a polymer is same as root mean square end-to-end distance of a polymer coil dissolved in theta solvent. Another method for determining root mean square end-to-end distance is the viscosity method of a dilute solution of a polymer in the theta solvent. The chain flexibility value is about 1.0 for flexible chains while for rigid chains it is more than 1.0.

8.7 Determination of ΔH_{mix}: Solubility Parameters

It is clear that solubility occurs when free energy of mixing G_{mix} is negative. In the case of non-polar molecules and in the absence of hydrogen bonding in molecules, only dispersion forces are present which makes the H_{mix} positive. H_{mix} can be expressed by

$$\Delta H_{mix} = V_m \left[\left(\frac{\Delta E_1}{V_1}\right)^{1/2} - \left(\frac{\Delta E^2}{V_2}\right)^{1/2}\right]^2 \phi_1\phi_2 \qquad ...(8.25)$$

where ΔH_{mix} = heat of mixing

V_m = total volume of the two liquids

ΔE_1 and ΔE_2 = energy of vaporization of liquids 1 and 2

V_1 and V_2 = molar volumes of the two liquids

ϕ_1 and ϕ_2 = volume fraction of the liquids 1 and 2

The term $(\Delta E/V)$ gives the amount of energy required to vaporize a unit volume of the liquid. We know energy is consumed during vaporization to overcome the forces of attraction, which holds the molecules in liquid state. Hence, this term $(\Delta E/V)$ is a measure of cohesive force or molar attraction constant that holds the molecules of liquid together and is given the name 'cohesive energy density' (CED). The square root of CED is solubility parameter. So that

$$\delta = \left(\frac{\Delta E}{V}\right)^{1/2} = \left(\frac{\rho\delta E}{M}\right)^{1/2} \qquad ...(8.26)$$

and equation (8.23) takes the form

$$\Delta H_{\text{mix}} = V_m(\delta_1 - \delta_2)^2 \phi_1\phi_2 \quad ...(8.27)$$

or $$\frac{\Delta H_{\text{mix}}}{V_m\phi_1\phi_2} = (\delta_1 - \delta_2)^2 \quad ...(8.28)$$

The factor $(\delta_1 - \delta_2)^2$ determines the magnitude of ΔH_{mix}. As the values of δ_1 and δ_2 tend to be equal, ΔH_{mix} becomes zero. In that case, Entropy is the sole factor responsible for the mixing to take place. Equation (8.27) can also be applied to polymer dissolution. The equation (8.27) will take the form

$$\Delta H = (\delta_1 - \delta_2)^2 \phi_p\phi_s \quad ...(8.29)$$

This equation tells us that, if, and are close to each other, then, polymer will undergo dissolution. But, if they are very different, then, polymer does not dissolve. This theory is valid for amorphous polymers and not for crystalline polymers. It cannot be applied to the solvents with polar groups and hydrogen bond. There should not be a positive interaction between the polymer and the solvent as the value of ΔH_{mix} becomes negative. In this case, high value of does not affect dissolution, and ΔH per unit volume can be calculated as follows (Hildebrand 1950).

The solubility parameter of solvents is measured from the change in molar internal energy during vaporization values. This method cannot be adopted for polymers as they dissociate even before the vaporization point is reached. It is interesting to note that the solubility parameter of polymer is same as the solubility parameter of the solvent in which it swells maximum.

Flory gave another method to determine ΔH_{mix} for a polymer solution. He presented the following equation

$$H_{\text{mix}} = \delta_s\delta_p \chi RT \quad ...(8.30)$$

where χ = Flory interaction parameter

This can have positive and negative values.

$$= Z\Delta WRT \quad ...(8.31)$$

where Z = the number of nearest neighbours (or coordination number). Small values of, whether positive or negative show spontaneity of dissolution.

8.8 Thermodynamics of Polymer Dissolution

The principles of thermodynamics do not vary with the size of the solute molecules. The enthalpy and entropy of the system vary during dissolution process. Hence, Gibbs equation tells if a solute will get dissolved/mixed in the solvent or not.

$$G = H - TS \qquad ...(8.32)$$

A homogeneous solution is obtained when Gibbs free energy of mixing $\Delta G_{mix} < 0$, i.e. Gibbs free energy of solution $G_{solution}$ is lower than the summation of Gibbs energy of solvent and solute $G_{solvent}$, G_{solute} respectively.

$$\Delta G_{mix} = G_{solution} - (G_{solvent} + G_{solute}) \qquad ...(8.33)$$

The value of ΔG_{mix} is affected by the following three factors:

(1) The heat released measured as ΔH on mixing polymer and solvent.

(2) The temperature T at which the process of dissolution is carried out.

(3) The increase of disorder due to mixing S_{mix}.

At constant temperature and pressure, it can be shown that

$$G_{mix} = \Delta H_{mix} - \mathrm{T}\,\Delta S_{mix} \qquad ...(8.34)$$

The process of dissolution is always accompanied by an increase in disorder so that ΔS is always positive. However, the temperature 'T' is always positive and Enthalpy or heat content (H) is minimum and the magnitude of change in entropy is lesser in the case of solution of a polymer than in the solution of low molecular weight solute. A reaction is spontaneous (whether chemical or physical) if there is a loss in enthalpy ($-\Delta H$) or gain in entropy ($+\Delta S$) and these two factors are independent of each other. During a reaction, three situations can arise: enthalpy would either increase ($+\Delta H$), decrease ($-\Delta H$) or remain unchanged ($H = 0$). Both the factors, i.e. a decrease in enthalpy and an increase in entropy ($-\Delta H$ and $+\Delta S$) support the change and the process becomes spontaneous. However, if there is an increase in enthalpy and a decrease in entropy ($+\Delta H$ and $-\Delta S$), then both the factors oppose the change with the result and the dissolution does not take place at all. The third situation is when one of the factors opposes and the other favours. In that case, the feasibility of reaction depends on the magnitude of the opposing factor. If the overall result shows a decrease in free energy then dissolution will occur. Thus, more the value of ΔH_{mix} is negative, higher would be the dissolution. Presence of attractive forces like hydrogen bonding, etc. between polymer and solvent makes ΔH_{mix} negative.

8.9 Flory-Huggins Theory of Polymer Solution

According to the theory of modified lattice model, every lattice site occupies the same volume. Each lattice space is occupied by either a solvent molecule or a chain segment. The polymer chains are divided into segments on the basis of ease of internal rotation, known as thermodynamic or Kuhn segments. The freely jointed chain is made up of segments of equal size which have free internal rotation. The lattice space occupied by a polymer segment would always need to have two more adjacent lattice spaces for the polymer molecule to maintain the continuity of the chain. It is for this reason that entropy of mixing is less in the case of polymer solutions than in the solution of small molecules. Because of this reason the polymer segments have lesser number of ways to be accommodated in lattice spaces. It is considered that polymer chain consists of covalently bonded segments whose size is of the same order as that of solvent molecules, i.e. $x = V_2 / V_1$ where Vs are the molar volume of components. In order to calculate the number of ways in which the polymer chain can be placed in a lattice, a restriction faced is that the segments occupy lattice sites continuously because of the presence of covalent bonding between the segments. The task is to find the mixing of N_1 solvent molecules with N_2 monodisperse polymer molecules that have segments. Suppose we add 'i' polymer molecules to an empty lattice having cells (lattice points). So for the first molecule

$$N_0 = (N_1 + xN_2) \qquad ...(8.35)$$

The number of vacant cells left, where the next molecule will be accommodated is given by $N_0 = x_i$ ...(8.36)

The first step is to find an expression for ΔS_{mix}^{combi} when $\Delta H_{mix} = 0$. We can use the equation

$$\Delta S_{mix}^{combi} = k[\ln\Omega_{12} - (\ln\Omega_1 + \ln\Omega_2)] \quad ...(8.37)$$

In the case of amorphous polymer that has > 1, each molecule of the polymer can adopt different conformation. The equation for this polymer solution becomes

$$\Delta S_{mix}^{combi} = -k\left[N_1 \ln\left\{\frac{N_1}{(N_1 + xN_2)}\right\} + N_2 \ln\left\{\frac{xN_2}{(N_1 + xN_2)}\right\}\right] \quad ...(8.38)$$

$$\text{or} \quad \Delta S_{mix}^{combi} = -R[\mathrm{n}_1 \ln\Phi_s + \mathrm{n}_2 \ln\Phi_p] \quad ...(8.39)$$

where Φ_s and Φ_p are the volume fractions of solvent and polymer respectively.

$$\Phi_s = \frac{N_1}{N_1 + xN_2} \text{ (8.35a) and } \Phi_p = \frac{xN_2}{N_1 + xN_2} \quad ...(8.40)$$

In terms of number of molecules n_1 and n_2 for solvent and polymer, the equation (8.39) takes the form

$$\Delta S_{mix}^{combi} = -R[n_1 \ln\Phi_1 + n_2 \ln\Phi_2] \quad ...(8.41)$$

This equation (8.41) can be compared with equation (8.37) for ideal mixing, where the molecule fraction is replaced by volume fraction. This is the general equation for athermal mixing. When $x = 1$ this equation is reduced to equation (8.37) for ideal solution.

The next parameter is Contact Gibbs free energy change $\Delta S_{mix}^{contact}$, which depends on the effects of intermolecular interactions. There are three types of contacts:

(1) solvent-solvent g_{11}

(2) segment-segment g_{22}

(3) solvent-segment Δg_{12}

A solvent-solvent and a segment-segment contact brakes to form two solvent-segment contacts. Hence, the Gibbs free energy change, to form a single solvent-segment contact represented by Δg_{12} can be calculated by the following equation

$$\Delta g_{12} = g_{12} - \frac{1}{2}(g_{11} + g_{22}) \quad ...(8.42)$$

If g_{12} is the total number of solvent-segment contacts in the solution then

$$\Delta G_{mix}^{contact} = (Z - 2)N_1\Phi_2\Delta g_{12} \qquad ...(8.43)$$

where Z is the number of lattice cells as first neighbours.

Since lattice parameter Z and Gibbs free energy change is not easily determined, a new parameter Δg_{12} is introduced known as Flory-Huggins polymer-solvent interaction parameter. It is defined as

$$\chi = (Z - 2)\frac{\Delta g_{12}}{kT} \qquad ...(8.44)$$

The value for $\Delta G_{mix}^{contact} = RTn_1\Phi_2\chi$ and the Gibbs free energy of mixing

$\Delta G_{mix} = \Delta G_{mix}^{contact} - T\Delta S_{mix}^{combi}$ leads to

$$\Delta G_{mix} = RT[n_1\ \Phi_1 + n_2 \ln \Phi_2 + n_1\Phi_2\chi] \qquad ...(8.45)$$

This equation given by Flory and Huggins gives the equilibrium thermodynamic properties of polymer solutions, e.g. large negative deviation from Raoult's law, phase separation, melting point depression in crystalline polymers, etc. This theory is able to explain the general trends but fails to give the result quantitatively. The major drawbacks in the theory are:

(1) The assumption of only one type of lattice space for solvent, polymer and mixture.

(2) It does not include self-intersection of chains.

(3) χ is not a simple parameter.

References

Barrett, K.E.J., *Dispersion in Organic Media*, Wiley, New York (1982)

Carothers, W.H., "An introduction to the General Theory of Condensation Polymers", *J. Am. Chem. Soc.*, 51, 2548-59 (1929).

Curran, D.P., Strategy Level Separation in Organic Synthesis: From Planning to Practice, *Angew. Chem. Int. Ed. Eng.*, 37. 1174-96 (1998).

Flory, P.J., *J. Chem. Phys.*, 10, 51 (1942).

Flory, P.J., *Principles of Polymer Chemistry*, Cornell Univ. Press, Ithaca, NY (1953).

H. Staudinger, 'Polymerization' *Ber. Dtsch. Chem. Ges.* B 53, 1073-85 (1920).

Hildebrand, J.H. and Scott, R.L., *The Solubility of Nonelectrolytes*, Van Nostrand-Reinhold, New York (1950).

Huggins, M.L., *Ann. N.Y. Acad. Sci.*, 43, 1 (1942).

——, *J. Am. Chem. Soc.*, 64, 1712 (1942).

——, *J. Phys. Chem.*, 45, 151 (1942).

Kuhn, W. "Über die Gestalt fadenförmiger Moleküle in Losungen" (On the shape of filform molecules in solutions). Kolloidzeitschrift 68, p. 2 (1934).

Natta, G., Pino, P., Corradini, P., Danusso, F., Mantica, E., Mazzanti, G. and Moraghio, G., *J. Am. Chem. Soc.*, 77, 1708 (1955).

Ziegler, K., Holzkamp, E., Breil, H. and Martin, H., *Agnew. Chem.*, 67, 541 (1955).

Appendix

Partial Molar Qualities and Chemical Potential

Extensive properties are those that change with the changes in the number of substances in the system, e.g. mass, total free energy of the system, enthalpy, entropy, etc. Intensive properties are that, which are independent of the changes in the number of substances in the system, e.g. temperature, density, molar free energy, molar entropy, etc. In a closed system, temperature and pressure are the factors that bring about the change of state. While in an open system that contains two or more components, the change in the number of moles of components plays a role in addition to the change in temperature and pressure, brings about a change of state. If an extensive property is represented by X, then X would be a function of temperature, pressure and number of moles of various constituents.

$$X = f\,(T, P, n_1, n_2 \ldots n_i) \qquad \ldots(1)$$

where $n_1 + n_2 + \ldots + n_i = N$ (total number of moles) for a small change in the number of moles of the ith components, the property dx is given by

$$dx = \left(\frac{\partial x}{\partial n_i}\right)_{T,\,P,\,n_1,\,n_2\ldots} dn_i \qquad \ldots(2)$$

This quantity $\left(\frac{\partial x}{\partial n_i}\right)_{T,\,P,\,n_1,\,n_2\ldots} dn_i$ is called the partial molar property represented by $\bar{x}_i$.

One of the most important partial molar properties is the partial molar free energy designated as chemical potential and is represented by:

$$\left(\frac{\partial G}{\partial n_i}\right)_{T,P,n_1,n_2\ldots} = \overline{G}_i = \mu_i \qquad \ldots(3)$$

We can define the chemical potential as the change in free energy of the system, which is obtained by addition of one mole of a component at constant pressure and temperature, to such a large quantity of the system that there is no appreciable change in the overall compositions of the system.

When a system is at equilibrium its properties are defined in terms of intensive properties.

The partial molar quantity in case of polymer solutions are chemical potential μ_i which can be defined in terms of partial molar Gibbs free energy $\overline{G_i}$ can be given as follows:

$$m_i = \left(\frac{\partial G}{\partial n_i}\right)_{T,P,n_i} \qquad \ldots(4)$$

for single component system, the partial molar qualities and corresponding molar quantities are same.

For a system at equilibrium, the chemical potential μ_i of component i is related to its activity a_i by the following relation

$$\mu_i - \mu_i^0 = RT\ln a_i \qquad \ldots(5)$$

where μ_i^0 = chemical potential of component i in its standard state

$$\mu_i - \mu_i^0 = \overline{\Delta G_i} \qquad \ldots(6)$$

the partial molar Gibbs free energy charge for the formation of the system which is

$$\mu_i = \left(\frac{\partial \Delta G}{\partial n_i}\right)_{T,P,n_i} \qquad \ldots(7)$$

Index